The Climate of History in a Planetary Age

The Climate of History

IN A PLANETARY AGE

Dipesh Chakrabarty

The University of Chicago Press   Chicago and London

The University of Chicago Press, Chicago 60637
The University of Chicago Press, Ltd., London
© 2021 by The University of Chicago
All rights reserved. No part of this book may be used or reproduced in any manner whatsoever without written permission, except in the case of brief quotations in critical articles and reviews. For more information, contact the University of Chicago Press, 1427 East 60th St., Chicago, IL 60637.
Published 2021
Printed in the United States of America

29 28 27 26 25 24 23 22 21 20   1 2 3 4 5

ISBN-13: 978-0-226-10050-0 (cloth)
ISBN-13: 978-0-226-73286-2 (paper)
ISBN-13: 978-0-226-73305-0 (e-book)
DOI: https://doi.org/10.7208/chicago/9780226733050.001.0001

Library of Congress Cataloging-in-Publication Data

Names: Chakrabarty, Dipesh, author. | Latour, Bruno.
Title: The climate of history in a planetary age / Dipesh Chakrabarty.
Description: Chicago ; London : The University of Chicago Press, 2021. | Includes bibliographical references and index.
Identifiers: LCCN 2020029264 | ISBN 9780226100500 (cloth) | ISBN 9780226732862 (paperback) | ISBN 9780226733050 (ebook)
Subjects: LCSH: Climatic changes—Social aspects. | Climatic changes—Political aspects. | Globalization. | Human ecology. | Civilization, Modern. | History—Philosophy.
Classification: LCC QC903 .C373 2020 | DDC 304.2/501—dc23
LC record available at https://lccn.loc.gov/2020029264

To Rochona and Arko

To the memory of those humans and other living beings who perished in the Australian firestorms of 2019–2020 and in the Amphan cyclone in the Bay of Bengal in 2020

Contents

Introduction: Intimations of the Planetary · 1

I * *The Globe and the Planet*

Chapter 1. Four Theses · 23
Chapter 2. Conjoined Histories · 49
Chapter 3. The Planet: A Humanist Category · 68

II * *The Difficulty of Being Modern*

Chapter 4. The Difficulty of Being Modern · 95
Chapter 5. Planetary Aspirations: Reading a Suicide in India · 114
Chapter 6. In the Ruins of an Enduring Fable · 133

III * *Facing the Planetary*

Chapter 7. Anthropocene Time · 155
Chapter 8. Toward an Anthropological Clearing · 182

Postscript: The Global Reveals the Planetary:
A Conversation with Bruno Latour · 205

Acknowledgments · 219
Notes · 227
Index · 277

* INTRODUCTION *

Intimations of the Planetary

For my part, these troubles move me neither to laughter nor again to tears, but rather to philosophizing, and to closer observation of human nature. For I do not think it right to laugh at nature, and far less to grieve over it, reflecting that men, like all else, are only a part of nature, and that I do not know how each part of nature harmonises with the whole, and how it coheres with other parts.

Baruch Spinoza to Henry Oldenburg (1675)[1]

If Hegel—a self-declared admirer of Spinoza—were alive to plumb the depths of our sense of the present, he would notice something imperceptibly but inexorably seeping into the everyday historical consciousness of those who consume their daily diet of news: an awareness of the planet and of its geobiological history. This is not happening everywhere at the same pace, for the global world remains undeniably uneven. The current pandemic, the rise of authoritarian, racist, and xenophobic regimes across the globe, and discussions of renewable energy, fossil fuel, climate change, extreme weather events, water shortage, loss of biodiversity, the Anthropocene, and so on, all signal to us, however vaguely, that something is amiss with our planet and that this may have to do with human actions. Geological events and events constituting the history of life until now have been the preserve of experts and specialists. But now the planet, however dimly sensed, is emerging as a matter of broad and deep human concern alongside our more familiar apprehensions about capitalism, injustice, and inequality. The COVID-19 pandemic is the most recent and tragic illustration of how the expanding and accelerating processes of globalization can trigger changes in the much longer-term history of life on the planet.[2] This book is about this emergent object-category of human concern, the planet, and how it affects our familiar stories of globalization. This conceptual shift has happened in my lifetime, and I hope I will be forgiven if I begin with a few autobiographical remarks.

Coming of age in the inegalitarian, turbulent, and left-leaning city of Calcutta in the 1960s, I grew up—like many other Indians of my generation—to value and desire an egalitarian and just social order. The enthusiasms of my adolescence later found an academic expression in my early work on labor history and in my association with the Indian *Subaltern Studies* project, which aimed to acknowledge the agency of socially subordinate people in the making of their own histories. Our thoughts were also profoundly influenced by the global rise of postcolonial, gender, cultural, minority, Indigenous, and other studies that the Australian academic Kenneth Ruthven gathered in the early 1990s under the rubric of "the new humanities."[3]

Caught up in the profound historical changes that the swirling currents of globalization had brought into the ordinary lives of middle-class Indians like me, I was at this time working as a historian and social theorist at the University of Melbourne. Even after my move to the University of Chicago in 1995, I remained preoccupied with questions that marked popular struggles in my youth, questions of rights, of modernity and freedom, and of a transition to a world more rational and democratic than the ones I had known. My book *Provincializing Europe: Postcolonial Thought and Historical Difference* (2000) was a product of these years in which I attempted, through a postcolonial frame, to develop means for understanding what anticolonial and modernizing elites in the formerly colonized countries did and could do, working sometimes at the limits of the imperial European intellectual legacies they inevitably inherited. This was what I could bring to the discussion of the story of the globe that European empires, anticolonial modernizers, and global capital had fashioned together, a theme that dominated history and other interpretive disciplines in the final decades of the last century and into this one.[4]

Something happened in the early part of this century that forced a shift in my own perspective. In 2003 a devastating bushfire in the Australian Capital Territory took some human lives as well as the lives of many nonhuman beings, gutted hundreds of houses, and destroyed all forests and parks that surround the famous "bush capital" of the country, Canberra. These were places I had grown to love while pursuing my doctoral studies there. The sense of bereavement occasioned by these tragic losses made me curious about the history of these particular fires and soon, as I read up on their causes, brought the news of anthropogenic climate change into the humanocentric thought world I used to inhabit. Scientists claimed that humans, in their billions and through their technology, had become a geophysical force capable of changing, with fearsome consequences, the climate system of the planet *as a whole*. I also learned of the burgeoning scientific literature on the Anthropocene hypothesis—the proposition that human impact on the planet was such as to require a change in the geological chronology of earth history to recognize that the planet had crossed the thresholds of the Holocene epoch (ca. 11,700 years old) and had entered an epoch deserving of a new name, the Anthropocene.[5]

The figure of the human had doubled, in effect, over the course of my lifetime. There was (and still is) the human of humanist histories—the human capable of struggling for equality and fairness among other humans while caring for the environment and certain forms of nonhuman life. And then there was this other human, the human as a geological agent, whose history could not be recounted from within purely humanocentric views (as most narratives of capitalism and globalization are). The use of the word *agency* in the expression "geological agency" was very different from the concept of "agency" that my historian-heroes of the 1960s—E. P. Thompson, for instance, or our teacher Ranajit Guha—had authored and celebrated. This agency was not autonomous and conscious, as it was in Thompson's or Guha's social histories, but that of an impersonal and unconscious geophysical force, the consequence of collective human activity.

The idea of anthropogenic and planetary climate change does not face much academic challenge these days, but the idea of the Anthropocene has been much debated by both scientists and humanist scholars.[6] The debate has also made the term into a popular and—as usually happens with such debates—a polysemic category in the humanities today. Yet whether or not geologists agree to formalize the label "Anthropocene" one day, the data amassed and analyzed over the last several years by the Working Group on the Anthropocene set up by the International Commission of Stratigraphy in London makes one thing clear: ours is not just a global age; we live on the cusp of the global and what may be called "the planetary."[7] In thinking of the last few centuries of human pasts and of human futures yet to come we need to orient ourselves to both what we have come to call the globe and to a new historical-philosophical entity called the planet. The latter is not the same as the globe, or the earth, or the world, the categories we have used so far to organize modern history. The intensification of capitalist globalization and the consequent crises of global warming, along with all the debates that have attended the studies of these phenomena, have ensured that the planet—or more properly, as I use it here, the Earth system—has swum into our ken even across the intellectual horizons of scholars in the humanities.

The globe, I argue, is a humanocentric construction; the planet, or the Earth system, decenters the human. The doubled figure of the human now requires us to think about how various forms of life, our own and others', may be caught up in historical processes that bring together the globe and the planet both as projected entities and as theoretical categories and thus mix the limited timescale over which modern humans and humanist historians contemplate history with the inhumanly vast timescales of deep history.

Capital, Technology, and the Planetary

The globe and the planet—as categories standing for the two narratives of globalization and global warming—are connected. What connects them are the phenomena of modern capitalism (using the term loosely) and technology, both global in their reach. After all, greenhouse gas emissions have increased almost exclusively through the pursuit of industrial and postindustrial forms of modernization and prosperity. No nation has ever spurned this model of development, whatever their criticisms of one another. As a result of the spread of industrialization, as historian John McNeill has pointed out, the twentieth century became "a time of extraordinary change" in human history. "The human population increased from 1.5 to 6 billions, the world's economy increased fifteen fold, energy use increased from thirteen to fourteen fold, freshwater use increased nine fold, and the irrigated areas by fivefold."[8] Given this global pursuit of industry and development, it is understandable that the proponents of climate justice should see global warming as a consequence of uneven capitalist development inflected by class, gender, and race and look suspiciously even on the topic of planetary climate change as an attempt to deny the less developed nations the "carbon space" they might need in order to industrialize.

Yet the history of capitalism alone, as it has been told until now, is not enough for us to make sense of the human situation today. This has to do with the dawning realization that many of today's "natural" disasters are consequences of changes that human socioeconomic institutions and technologies cause in processes that Earth system scientists regard as planetary. These processes until now have operated mostly independently of human activities but have nevertheless been central to the flourishing of human and other forms of life. The more we acknowledge our emerging planetary agency, the clearer it is that we now have to think about aspects of the planet that humans normally just take for granted as they go about the business of their everyday lives. Take the case of the atmosphere and the share of oxygen in it. The atmosphere is as fundamental to our existence as the simple act of breathing. But what is the history of this atmosphere? Do we need to think about that history today in thinking about human futures? Yes, we do. For the last 375 million years—since the evolution of large forests, that is—the concentration of oxygen has been maintained by certain processes on the planet at a level that ensured that animals did not suffocate from lack of oxygen and forests did not burn from an overabundance of it. Diverse dynamic processes maintain the atmosphere in its current equilibrium. Oxygen being a reactive gas, the air needs a constant supply of fresh oxygen. Some of this oxygen comes even from tiny sea creatures like plankton. If human activities affecting the sea completely destroyed such plankton, we would thereby destroy a major source of oxygen. In short, humans have acquired the capacity to interfere with planetary processes but not necessarily—at least not as yet—the capacity to fix them.

Our abilities to shape the planet are largely technological, so technology also is an intrinsic part of this unfolding story about humans. The geologist Peter Haff recently introduced the concept of the "technosphere" to characterize the global system of human technology:

> The proliferation of technology across the globe defines the technosphere—the set of large-scale networked technologies that underlie and make possible rapid extraction from the Earth of large quantities of free energy and subsequent power generation, long-distance, nearly instantaneous communication, rapid long-distance energy and mass transport, the existence of modern governmental and other bureaucracies, high-intensity industrial and manufacturing operations including regional, continental and global distribution of food and other goods, and myriad additional "artificial" or "non-natural" processes without which modern civilization and its present 7×10^9 human constituents could not exist.[9]

According to Haff's argument, the human population at its current size is "deeply dependent on the existence of the technosphere" without which it "would quickly decline towards its Stone Age base of no more than ten million . . . individuals."[10] Technology, one could then say with Haff, has become a condition for biology, for the existence of humans in such massive numbers on the planet.[11]

Haff's thesis about the technosphere enables us to see how "unencumbered," in Carl Schmitt's terms, technology has become today—and how, given the power of technology, humans have already made Earth into a spaceship for themselves and other forms of life that depend for their own existence on human flourishing. In his 1958 "Dialogue on New Space," Schmitt articulated through the voice of a fictional character, Mr. Altman (an old historian), a fundamental distinction between living on land and living on a ship at sea. At the core of "a terrestrial existence," he said, there stood "house and property, marriage, family and hereditary right" along with domestic and other animals. Technology, when present in this kind of life, would be encumbered with all that such a life entailed. Technology per se would never be in charge of this life. With the conquest of the seas, however, the ship came to embody what Schmitt called "unencumbered technology." Unlike the house of "terrestrial existence," at the core of "maritime existence" was the ship, a "much more intensely . . . technological means than the house." On the ship (as on airplanes today), life is crucially dependent on the proper functioning of technology.[12] If technology fails, life faces disaster. If Haff's argument is right that the technosphere today has become the primary condition for the survival of seven (soon to be nine) billion humans, we could say that we have already made Earth into something like Schmitt's ship in that its capacity to support our many billion lives is now dependent on the existence of the technosphere itself. In a later article, where Haff distinguishes between a "social Anthropocene"—one "that engages the conditions, motivations and histories of the world's peoples, including the role of politics"—and a "geological Anthropocene," he reiterates that it is important for humans "to recognise that the technosphere has agency, and that agency is not the same as our own."[13]

The technosphere extends deeply into "subterranean rock mass via mines, boreholes and other underground constructions" and into the "marine realm" as well—not only through ships and submarines but also through "oil platforms and pipelines, piers, docks, [and] aquaculture structures."[14] On dry land, it encompasses our "houses, factories, and farms" along with our "computer systems, smartphones, and CDs" and "the waste in landfills and spoil heaps." The technosphere "is staggering in scale, with some thirty trillion tons representing a mass of more than 50 kilos for every square metre of the Earth's surface." "The technosphere," observes geologist Mark Williams, "can be said to have budded off the biosphere and arguably is now at least partly parasitic on it." And compared to the biosphere, "it is remarkably poor at recycling its own materials, as our burgeoning landfill sites show."[15]

Equally striking are the figures illustrating the role humans have played in reshaping the landscape of the planet not only on its surface but down to the continental shelves. The planet's land and seafloor have been transformed by humans. "By the end of the 20th century, sea-bottom trawling was taking place across an area of some 15 million km^2 each year. This now includes most of the world's continental shelves and significant areas of the upper continental slope, along with the upper surfaces of seamounts."[16] In 1994, according to one estimate, "human earth moving caused 30 billion tons to be moved per year on a global basis." A 2001 estimate gives the figure of 57 billion tons per year. For comparison, the amount of sedimentation carried into the ocean by the world's rivers each year amounts to between 8.3 and 51.1 billion tons a year.[17] Humans, say the geologist Colin Waters and his colleagues, "now move more sediment around in this fashion [mining and quarrying] than all natural processes combined (26 Gt/yr)."[18] Such a considerable geomorphological and biological role cannot be separated from the history that connects capitalism with global warming.

If all this and much else about human impact on the planet suggest to Earth system scientists that the planet may have passed the threshold of the Holocene and entered a new geological epoch altogether, we can then say that as humans we presently live in two different kinds of "now-time" (or what they call *Jetztzeit* in German) simultaneously: in our own awareness of ourselves, the "now" of human history has become entangled with the long "now" of geological and biological timescales, something that has never happened before in the history of humanity.[19] True, earth-scale phenomena—earthquakes, for instance—erupted into our humanist narratives, no doubt, but for the most part geological events such as the uplift or erosion of a mountain occurred so gradually that mountains were seen as a constant and unchanging background to human stories. In our own lifetime, however, we have become aware that the background is no longer just a background. We are part of it, acting as a geological force and contributing to the loss of biodiversity that may, in a few hundred years, become the sixth great extinction. Irrespective of whether the term is ever formalized or not, the Anthropocene signifies the extent and the duration of our species' modification of the earth's geology, chemistry, and biology.[20]

In thinking historically about humans in an age when intensive capitalist globalization has given rise to the threat of global warming and mass extinction, we need to bring together conceptual categories that we have usually treated in the past as separate and virtually unconnected. We need to connect deep and recorded histories and put geological time and the biological time of evolution in conversation with the time of human history and experience. And this means telling the story of human empires—of colonial, racial, and gendered oppressions—in tandem with the larger story of how a particular biological species, *Homo sapiens*, its technosphere, and other species that co-evolved with or were dependent on *Homo sapiens* came to dominate the biosphere, lithosphere, and the atmosphere of this planet. We have to do all this, moreover, without ever taking our eyes off the individual human who continues to negotiate his or her own phenomenological and everyday experience of life, death, and the world—experience that takes for granted a "world" that today, ironically, no longer presents itself as simply given.[21] The crisis at the planetary level percolates into our everyday life in mediated forms and, one could argue, it even issues in part from decisions we make in everyday life (such as flying, eating meat, or using fossil-fuel energy in other ways). But that does not mean that the human phenomenological experience of the world is over. True, we are never distant from deep time and deep history. They run through our bodies and lives. Humans in everyday lives may be forgetful of their evolved characteristics, but the design of all human artifacts, for instance, will always be based on the assumption that humans have binocular vision and opposable thumbs. Having big and complex brains may very well mean that our big and deep histories can exist alongside and through our small and shallow pasts, that our internal sense of time—that phenomenologists study, for instance—will not always align itself with evolutionary or geological chronologies.[22]

Being Political at the Limits of the Political

The coming together of human and nonhuman scales produces the political in the form of a paradox that calls into question previous ways of thinking about and using that category.[23] My use of the word *political* is indebted to Hannah Arendt's thoughts modified by my reading of Carl Schmitt. The innate connection that exists between intergenerational time and Arendt's conception of the political allows us see why *any* action undertaken with the aim of addressing climate change over a time span covering the lives of multiple generations is political (though no one solution will be to everybody's satisfaction).[24]

Readers of *The Human Condition* will remember that Arendt identified the human capacity for making use of individual differences—in her terms, *plurality*—to create the new or the novel in human affairs as the source of "action." "Action" is foundational to her definition of the political. Action, Arendt wrote, "corresponds to the human condition of plurality, to the fact that men, not Man, live on earth and inhabit the world."[25] Action is "the political activity par excellence." Action was also tied to the condition of natality—the fact that we are all born as new and unique individuals. "Action in so far as it engages in founding and preserving political bodies," writes Arendt, "has the closest connection with the human condition of natality; the new beginning inherent in birth can make itself felt in the world only because the newcomer possesses the capacity of beginning something anew, that is, of acting."[26] The possibility of newness, that is, natality—"and not mortality," adds Arendt—remains "the central category of political, as distinguished from metaphysical, thought."[27] Arendt returned to the idea of natality in her later publication *The Life of the Mind*: "Every man, being created in the singular [unlike animals or species-being, says Arendt], is a new beginning by virtue of his birth."[28] The point is repeated again in *The Promise of Politics*: "*man* is apolitical. Politics arises *between men*, and so quite *outside of man*. There is therefore no political substance. Politics arises in what lies *between men* and is established as relationships."[29]

Arendt's ideas about the political have sometimes been criticized for their apparent lack of interest in relations of domination, injustice, inequality, and, by extension, democracy.[30] But her conception of "the political" can be put in conversation with her ideas about "action" and "work" to create a conceptual space that allows for an interest in precisely the issues that Arendt's critics thought she disdained.[31] The triple distinction between labor, work, and action with which Arendt opens *The Human Condition* enables us to see the point more clearly.[32] "The human condition of labor is life itself," writes Arendt. It is literally about consumption—the metabolism we need to sustain our biological bodies and their eventual and inevitable decay. What labor sustains—the individual body—does not live beyond the individual's lifetime. In contrast, "work" has to do with all kinds of human artifice—from language and institutions to man-made things—that are necessarily *intergenerational*.[33] Work produces "the world of things." Though "each individual life is housed" in this "world," the "world itself is meant to outlast and transcend them all." Work thus produces intergenerational time as constitutive of itself. This idea of intergenerational time is encapsulated in the argument that work generates matters that endure, though admittedly, usage "wears out" their durability.[34] The world that precedes us in time and yet leaves to us its enduring institutions, ideas, practices, and things has to be intergenerational in orientation. Arendt connects this with the idea of dwelling: "In order to be what the world is always meant to be, a home for men during their life on earth, the human artifice must be a place fit for action and speech."[35] For artifices to act as this site, they need to survive the logics of pure consumption and utility.[36]

Political action, in this sense, is that which helps humans to be at home on earth beyond the time of the living. A consumption-driven capitalism in which all artifacts are up for consumption in the present would be an antipolitical machine in that it would eventually work against the logic of human dwelling, since dwelling requires artifacts to endure beyond the lifetime of the living. It would be akin to Arendt's category "labor"—the activity that all animals have to engage in, which is finding food to keep biological life going. Intergenerational concerns, made howsoever difficult by the fact that the unborn are not there to press their claims against those of the living, are thus central to Arendt's conception of the political.[37] Questions of climate justice—not only between the rich and the poor but also between the living and the unborn—surely fall under the political in this view. How humans might transition to renewable energy, develop sustainable societies, and other such questions of concern would also be, by the same token, political. Needless to say, the word *political*, thus applied, would refer to *all* activities undertaken to deal with the consequences of—and hence the future posed by—global warming, from scientific, technological, and geoengineering experiments to policy work and activism across the spectrum of all available ideologies.

Armed with this conception of the political (I will soon add a Schmittian modification to it), how should we conceive of our own times as we add to the postcolonial, postimperial, and global concerns of the last century issues such as anthropogenic climate change and the Anthropocene? The emergence of these latter issues surely does not mean that the issues that seemed important in the postcolonial world and in the context of globalization have gone away. After all, we still live in times when representation of histories of "people without history" remains a debated question, when the question of sovereignty of those who lost their lands and civilizations to European occupiers and invaders remains unanswered (and perhaps, disquietingly, unanswerable), when inequalities between classes grow more acute and wealth concentrates in the hands of the so-called one percent, and when the number of refugees or stateless people in the world keeps swelling while global capital pursues technology that drastically changes and threatens the future of human labor. The same digital technology that makes for intelligent machines also acquires a Janus-faced presence in the lives of democracies: social media applications like WhatsApp and Facebook can help popular mobilization but are not necessarily conducive to the nuanced debates and discussions that democratic deliberations also require.

To talk about the planetary and the Anthropocene is not to deny these problems but to render them layered in both figurative and real terms. The geological time of the Anthropocene and the time of our everyday lives in the shadow of global capital are intertwined. The geological runs through and exceeds human-historical time. Some consequences of human impact on the planet—cities becoming heat islands, rise in the fre-

FIGURE 1. Theo, age two

quency and intensity of hurricanes, acidification of the seas—remain visible in historical time. Others—like the impact anthropogenic climate change may have on the glacial-interglacial cycle that has characterized the history of this planet for more than two million years—are not. Some of the results of our capacity to move earth around are visible and often ugly. In the state of Rajasthan in India, thirty-one hills that have "gone missing"—that is to say razed to the ground illegally by criminal businessmen looking for "raw materials" to feed the construction boom in the country—are an ugly demonstration of the earth-moving capacity of modern humans and their machinery.[38] But when I see in a neighborhood park a child unselfconsciously walking around an earthmoving machine and then see the same child moving sand in a sand pit with the help of miniature versions of the same machinery— Anthropocene toys!—I see how much our geomorphological agency has been "naturalized" (figs. 1, 2). There is no question of artificially separating the time of the Anthropocene from the human time of our lives and history. In many ways, our capacity to act as a geophysical force is connected to many modern forms of enjoyment.

If anything, many of the problems we identify as problems of capital-

FIGURE 2. Theo and friends

ist globalization will only intensify as global warming increases. Shrinking habitability of the planet, a rise in the number of climate refugees and "illegal immigrants," water scarcity, frequent extreme weather events, prospects of geoengineering, and so forth cannot be a recipe for global peace.[39] In addition, our global failure to create a governance mechanism for planetary climate change also suggests that we are not dealing with the kind of "global" problems our global governance apparatus, the United Nations, was set up to deal with.

There is an interesting problem of temporality here. Negotiations between nations at the level of the UN usually assume an open and indefinite calendar. For instance, we don't know when there will be peace between the state of Israel and the Palestinian population or whether the people of Kashmir will ever live in an undivided land. Those are questions that belong to an open and indefinite calendar. Similarly, we don't know when humans will be successful in ushering in a fair and equitable world. The struggle against capitalism assumes that there is time aplenty for our historical questions of injustice to be settled. The climate problem and all talk of "dangerous" climate change, on the other hand, confront us with finite calendars of urgent action. Yet powerful nations of the world have sought to deal with the problem with an apparatus that was meant for actions on indefinite calendars. Following the success of the Montreal Protocol of 1987, the UN treated anthropogenic climate change as a "global"—and not planetary—problem that was to be resolved through the UN mechanism. This is why the UN set up the Intergovernmental Panel on Climate Change (IPCC) in 1988. But, interestingly, the action that the IPCC recommends—on global carbon budgets, for instance—assumes a finite and definite calendar that then is subjected to global negotiations. The 2°C figure that is normally seen as the threshold of "dangerous" climate change, for instance, represents a politically negotiated compromise between the UN tendency toward an indefinite calendar of action and the finite calendar that scientists come up with. It is entirely possible that planetary climate change is a problem that the UN was not set up to deal with. But we have no better democratic alternative at present. Climate change and the Anthropocene are thus problems that are profoundly political and that challenge our received political institutions and imaginations at the same time.[40]

While being guided by Arendt's ideas about the political, it is important not to lose sight of the Schmittian insight that even if humans are capable of being rational and creative, there is no humanity that can act as the bearer of a single, rational consensus. "The political world," writes Schmitt, "is a pluriverse, and not a universe."[41] One does not know where and how human history will proceed. Our times also require us to address another point that neither Schmitt nor Arendt ever addressed. They are both helpful in giving us a capacious understanding of the political, but because this understanding remains focused only on humans, it is unfortunately not capacious enough. Animals and other nonhumans cannot be part of the political in Arendt's or Schmitt's schema. Yet the planetary environmental crisis calls on us to extend ideas of politics and justice to the nonhuman, including both the living and the nonliving. The more this realization sinks in, the more we realize how irrevocably humanocentric all our political institutions and concepts are. The important point is that the climate crisis and the Anthropocene hypothesis together represent an intellectual and political quandary for humans and warrant new interpretations of the significance and meanings of what I once called "political modernity."[42] Hence, the work of pioneering thinkers such as Bruno Latour, Isabelle Stengers, Donna Haraway, Jane Bennett, and others who have long studied this question of extending the political beyond the human is relevant and important to this project. I will have comments to make on this problem without in any way claiming to have solved it.

As should be clear by now, it does not matter for my argument whether or not the Anthropocene label is one day accepted formally by geologists as the name of our current geological epoch. As Zalasiewicz and his colleagues say, the "future status" of the term "as a concept" can "in general be regarded as secure but is uncertain in formal terms."[43]

Jeremy Davies, Eva Horn, Hannes Bergthaller, and others are right to say that for humanists, the main benefit of the discussion about the Anthropocene is that it has brought the geobiological into view. My particular concern has been to figure out what the implications of the science of climate change and Earth System Science could be for humanists interested in thinking about the historical time through which we are passing. I should clarify, however, that I do not approach science from within the traditions developed in certain branches of science studies that often make, in Bernard Williams's words, "the remarkable assumption that the sociology of knowledge is in a better position to deliver the truth about science than science is to deliver the truth about the world."[44] I consider it indisputable that the pursuit of science remains enmeshed in the politics of class, gender, race, economic regimes, and scientific institutions. Therefore, concerns about the actual power and authority particular scientists may exercise in particular historical contexts are entirely legitimate. I do not believe, however, that such enmeshment makes the *findings* of scientific disciplines any more arbitrary or false or merely political than empirical statements and analyses by a fellow historian or social scientist.[45] Without the sciences, there still would have been atmospheric warming and erratic weather, but there would not have been an intellectual problem called "planetary climate change" or "global warming" or even the Anthropocene. This is not to deny the need to produce two-way, practical translations in places between local knowledge, customs, traditions, practices, and a science that is planetary in scope but to plainly acknowledge that "the local," by itself, would have never given us any understanding of the roles that parts of the world sparsely or not at all inhabited by humans—such the regions containing the Siberian permafrost or the oceans themselves—play in processes that determine cooling or warming of the whole planet.[46]

The Evolution and Structure of This Book

Troubled by my own thoughts about planetary climate change but also stimulated by the methodological challenges the problem of geological agency of humans posed to my usual habits of thought as a humanist historian, I published an essay in 2009—"The Climate of History: Four Theses." I wondered about what humanity was in this age of the Anthropocene. We are simultaneously a divided humanity as well as a dominant partner in a techno-socioeconomic-biological complex that includes other species. It is this complex that is driving species extinction and is thus itself a part of the history of life on this planet. This makes the complex a geological agent as well. With the collapsing of multiple chronologies—of species history and geological times within living memory—the human condition has changed. This changed condition does not mean that the related but different stories of humans as a divided humanity, as a species, and as a geological agent have all fused into one big geostory and that a single story of the planet and of the history of life on it can now serve in the place of humanist histories. As humans, I argued, we have no way of experiencing in unmediated forms these other modes of being human that we know cognitively at an abstract level. Humans in their internally differentiated plurality, humans as a species, and humans as the makers of the Anthropocene constitute three connected but analytically distinct categories. We construct their archives differently and employ different kinds of training, research skills, tools, and analytical strategies to conceive of them as historical agents, and they are agents of very different kinds.[47]

This conscious conjoining of differently scaled chronologies produced in me a feeling that I have often likened to that of falling. It was as though as a humanist historian interested in political issues of rights, justice, and democracy, I had fallen into "deep" history, into the abyss of deep geological time. This falling into "deep" history carries with it a certain shock of recognition of the otherness of the planet and its very large-scale spatial and temporal processes of which humans have, unintentionally, become a part. Being from the Indian subcontinent, where diabetes has acquired epidemic proportions, I sometime explain this experience by drawing an analogy with how an Indian person's sense of his or her own pasts suddenly undergoes an instantaneous expansion when he or she is diagnosed as diabetic. You go to the doctor with a (potentially) historian's view of your own pasts, a biography that you could place in certain social and historical contexts and that spans two or three generations. The diagnosis, however, opens up completely new, impersonal, and long-term pasts that could not be owned by one in the possessive-individualist sense that the political theorist C. B. Macpherson once wrote about brilliantly.[48] A subcontinental person will most likely be told that they have a genetic propensity toward diabetes because they have been rice eaters (for at least a few thousand years now). If they were academic and from a Brahmin or upper-caste family in addition, then they had practiced a sedentary lifestyle for at least a few hundred years, and it would perhaps also be explained to them that human muscles' capacity for retaining and releasing sugar was related to the fact of humans having been hunters and gatherers for the overwhelming majority of their history—suddenly, evolution and deep history![49] You don't have experiential access to any of these longer histories, but you fall into a sudden awareness of them.

While my 2009 essay received many appreciative responses, it also confronted a maelstrom of criticism. My critics claimed that my reference to humans as a "dominant species" and my use of the moniker Anthropocene (and not something like Capitalocene) ran the risk of "depoliticizing" climate change by detracting attention from questions of responsibility, the role of capitalism, empires, uneven development, and the drive for capitalist accumulation. The rich, they rightly pointed out, were far more responsible for the climate crisis and would always be lesser victims of the crisis than the poor. I have replied to some of the specific criticisms elsewhere.[50] Without going over that ground in detail, let me simply say that I agree with the literary scholar Gillen D'Arcy Wood's recent assessment that some of the criticism may have been at cross purposes.[51]

However, it may be productive to reflect on the burden of the criticisms I received for what it tells us about the recent history of the interpretive disciplines in the human sciences. Scholars in the field of postcolonial history and theory were relatively slow to respond to the crisis of global warming even though the science was hitting the newsstands by the late 1980s. Take the year 1988. That was when James Hansen, then the head of the Goddard Space Center of NASA, spoke to the United States Senate and presented them with three main conclusions: "Number one, the earth is warmer in 1988 than at any time in the history of instrumental measurements. Number two, the global warming is now large enough that we can ascribe with a high degree of confidence a cause and effect relationship to the greenhouse effect. And number three, our computer climate simulations indicate that the greenhouse effect is already large enough to begin to [affect] the probability of extreme events such as summer heat waves."[52] The United Nations' Intergovernmental Panel on Climate Change (IPCC) was set up the same year.

Postcolonial thinking and criticism—a branch of the humanities that deeply influenced critical theory in the late twentieth century—may be said to have begun its journey ten years before these events with the publication in 1978 of Edward Said's polemical classic *Orientalism*.[53] Many currents of thought—critical race theory, feminist criticism, anticolonial and postcolonial criticism, cultural studies, minority studies—came together in the 1980s and after to move the humanities away from the aristocratic temperament in which the study of texts, rhetoric, philology, grammar, prosody, and so forth had been anchored for ages.[54] Humanities, a branch of knowledge that once was about the cultivation of personhood and establishing cultural claims to rule by imperial and other elites, transformed in these decades into a branch of knowledge that both studied and produced what James Scott once memorably called "weapons of the weak."⁵⁵ *Subaltern Studies*, first published in 1983, was conceived very much within this new orientation of the humanities.⁵⁶ The year 1988 was when Gayatri Chakravorty Spivak published her famous essay "Can the Subaltern Speak?"⁵⁷ And within a few years from of the publication of that essay, Homi Bhabha, Stuart Hall, Kobena Mercer, and others collaborated in 1995 to curate the first postcolonial exhibition and conference on Franz Fanon that resulted in the publication of *The Fact of Blackness*.⁵⁸ Bhabha's own classic collection of essays on questions of postcolonial criticism and thinking, *The Location of Culture*, came out in 1994.⁵⁹

These new intellectual directions in the humanities produced revelatory insights but remained, if I may say so without putting too fine a point on it, environmentally blind. This might sound like an extreme statement, so let me briefly explain what I mean. The late 1960s and the 1970s had, of course, seen an upsurge of environmentalist movements in different parts of the world. Europe saw the rise of green parties and politics. In many cases environmentalist movements and thoughts led to certain criticisms of capitalist growth models (as may be seen in the writings of the well-known Indian environmentalist Vandana Shiva). The tocsin that Rachel Carson sounded with her 1962 *Silent Spring* led, thanks in no small measure to the "blue marble" picture of earth popularized by American astronauts in 1972, to the idea that humans had this "one world"—one atmosphere, one large mass of oceanic water, one spherical home—that was both fragile and finite, vulnerable to the ravages of extractive capitalism and the postindustrial way of life. One of the most prominent expressions of this "finite earth" view was the 1972 report called *The Limits to Growth*, also known as the Club of Rome Report authored by Donella and Dennis Meadows and their colleagues.⁶⁰

But this "one-worldism" did not quite gel with the humanities traditions from which postcolonial criticism issued. Scholars in the humanities (in which I include the interpretive branches of history and anthropology) were fundamentally not "lumpers" but "splitters" of human history, scholars who believed that all claims about the "oneness" of the world had to be radically interrogated by testing them against the reality of all that actually divided humans and formed the basis of different regimes of oppression: colony, race, class, gender, sexuality, ideologies, interests, and so on. They were skeptical of arguments that tended to fold the diversities of human worlds into the oneness of a finite Earth. This unifying move seemed ideologically suspect and always appeared to have been made in the interests of power. These scholars believed that the path for the emancipation of all humans one day could not be found without *first* addressing and working through the conflicts and injustices these divisions entailed. Indeed, one may see this opposition between the lumpers/one-worlders and the splitters/postcolonials as running through many of the legitimate demands that are made in the name of environmental or climate justice today.[61] The splitting reflex is now deeply set and thoroughly understandable: scholars in the humanities, after all, have been raised—and with good reason—for over five decades to be extremely suspicious of all claims of totality and universalism. I was myself a child of this tradition.

Given this tension between "splitters" and "lumpers" that my own thinking was also once subject to, I spent the last ten years trying to work my way through the implications of—and the controversies surrounding—the four theses I set out in the essay of that name. This is why, much discussed and critiqued though it has been, that essay, "The Climate of History: Four Theses," now revised and renamed "Four Theses," remains the inescapable starting point of this project and serves as the first chapter of this book. I have added a brief appendix to this chapter to clarify my use of words and expressions such as *species* and "negative universal history" that troubled some readers of the original essay on which this chapter is based.

Methodologically, this book presents an argument by taking the reader through the path I myself took to develop and arrive at it. A key moment along that path of discovery came for me when I stumbled on the realization that the concept of *globe* in the word *globalization* was not the same as the concept of *globe* in the expression *global warming*. Same word but their referents were different. This is the idea that I would eventually come to develop as the globe/planet distinction. Not a binary, I wish to stress, but two related terms distinguished analytically. The more I read into Earth System Science and the more I thought about the accusation that my four theses let capitalism off the hook, the more this distinction seemed important to me. This is what the first section of the book is all about. My second chapter, on "Conjoined Histories," where I put out some initial thoughts about why the few-centuries-old story of capitalism did not give us enough of an intellectual grip on the problems of human history that anthropogenic climate change revealed, builds up to the globe/planet distinction. It is in chapter 3, "The Planet: A Humanist Category" that the distinction is fully developed. This chapter provides the intellectual fulcrum on which the argument of this book turns. The category "planet" allowed to me see, and ultimately to say, that contemplating our own times required us to behold ourselves from two perspectives at once: the planetary and the global. The global is a humanocentric construction; the planet decenters the human.

The second section of the book, titled "The Difficulty of Being Modern," consists of three chapters that variously explore the question of why modern ideas of freedom—whether projected for individuals, the nation, or for humanity in general—retain their attraction even after many of the assumptions underlying them have been justifiably challenged by various critics of modernity and modernization. In the first of these (chap. 4), "The Difficulty of Being Modern," I analyze the intimate connection between postcolonial nations' conceptions of freedom and the increased need for energy that has been historically serviced mainly by fossil fuel and by a variety of projects of "mastery of nature" (such as damming rivers). The second (chap. 5), "Planetary Aspirations: Reading a Suicide in India," is a reading of the ideas of humanity and freedom that a Dalit-identified young man who took his life in 2016 left in his suicide note. The chapter reads the history of stigma and upper-caste disgust surrounding "the dalit body" as pointing to certain limits to how the human body is imagined in the reigning conceptions of the political. The last chapter (chap. 6) in this section, a reading of Kant's 1786 essay "Speculative Beginning of Human History," is in effect an anthropocenic critique trying to show how the distinction, fundamental to modernity, made by the great philosopher between the moral and animal lives of humans has come undone in the present crisis of the biosphere.

I have called the final section of the book "Facing the Planetary," partly as a homage to the political theorist William Connolly's book by the same name. It begins with a chapter called "Anthropocene Time"—chapter 7—where I try to explicate what the geologist Jan Zalasiewicz calls a planet-centered mode of thinking in order to distinguish it from thoughts that put human concerns exclusively at their center. I then go on to develop—in a chapter titled "Towards an Anthropological Clearing" (chap. 8)—some of the implications that our recent opening up to the planetary and the geological has for our understanding of the human condition today. I look on the present crisis as providing an opportunity for working toward Karl Jaspers's idea of an "epochal consciousness," a form of argumentation that seeks to make a conceptual place for thinking the human condition *before* committing to any particular version of practical or activist politics. The fact of the planet—the category explained in the first section—coming into view in the everyday lives of humans leads us to question whether the relationship of mutuality between humans and the earth/world that many twentieth-century thinkers inherited, assumed, and celebrated has become untenable today. How do we move, in the face of the current ecological crisis, toward composing a new "commons," a new anthropology, as it were, in search of a redefinition of human relationships to the nonhuman, including the planet? That is where this book ends, with the beginnings, I hope, of a conclusion not yet reached in history.

A postscript of a conversation with Bruno Latour is added in which many of the points made in this book are canvassed.

As the foregoing description of the book makes obvious, my account of human worlds and their relationship to the planet humans inhabit does not aim to contribute in any immediate and practical sense to possible solutions to climate-related conflicts in the world. These conflicts, as I have said, may even be sharpened by the planetary environmental crisis and by the different tensions—having to do with borders, water, food, housing—that the crisis provokes. My hope is that stories that bring our deep and recorded histories together may help generate new perspectives from which to view these conflicts and thus contribute indirectly toward their mitigation. The more we see that in spite of all our divisions and inequalities, what is at stake is the survival of civilization as we have known it, the more, I hope, we will see the insufficiency of our necessarily partisan views in addressing what one may, with a nod toward Arendt, call the human condition today.

In his book on the Anthropocene, philosopher Sverre Raffnsøe reminds us that Kant's *Introduction to Logic* saw four fundamental questions as critical to distinguishing four domains of knowledge: the basic question of "knowledge, science, theory and metaphysics" was "What can I know?"; the basic question of moral and practical thought, "What should I do?" and the question in religious and aesthetic contexts was "What can I hope for?" Raffnsøe goes on to point out that Kant saw these three basic questions as "folding in on each other, and mutually contributing to and being clarified through . . . a fourth and seminal question: 'What is Man? [*Was ist der Mensch?*].'"[62] The last question, wrote Kant, was to be answered by "anthropology."[63] The phenomenon of the Anthropocene and the crisis of climate change raise all of these questions. Scientists—and social scientists—are best placed to discover what we can know; politicians, activists, and policy thinkers and engineers are best equipped to find what we can do. It falls to religion, aesthetic, and cognate domains of thought to suggest what we can expect. Humanity's current predicament renews for the humanist the question of the human condition. This humble book, then, joins the efforts of other humanists in collectively thinking our way toward a new philosophical anthropology.

PART I
✶
The Globe and the Planet

1
Four Theses

The planetary crisis of climate change or global warming elicits a variety of responses in individuals, groups, and governments ranging from denial, disconnect, and indifference to a spirit of engagement and activism of varying kinds and degrees. These responses saturate our sense of the now. Alan Weisman's best-selling book *The World without Us* suggests a thought experiment as a way of experiencing our present: "Suppose that the worst has happened. Human extinction is a fait accompli. . . . Picture a world from which we all suddenly vanished. . . . Might we have left some faint, enduring mark on the universe? . . . Is it possible that, instead of heaving a huge biological sigh of relief, the world without us would miss us?"[1] I am drawn to Weisman's experiment as it tellingly demonstrates how the current crisis can precipitate a sense of the present that disconnects the future from the past by putting such a future beyond the grasp of historical sensibility. The discipline of history exists on the assumption that our past, present, and future are connected by a certain continuity of human experience. We normally envisage the future with the help of the same faculty that allows us to picture the past. Weisman's thought experiment illustrates the historicist paradox that inhabits contemporary moods of anxiety and concern about the finitude of humanity. To go along with Weisman's experiment, we have to insert ourselves into a future "without us" in order to be able to visualize it. Thus, our usual historical practices for visualizing times, past and future, times inaccessible to us personally—the exercise of historical understanding—are thrown into a deep contradiction and confusion. Weisman's experiment indicates how such confusion follows from our contemporary sense of the present insofar as that present gives rise to concerns about our future. Our historical sense of the present, in Weisman's version, has thus become deeply destructive of our general sense of history. By history, of course, I refer here to the humanist art of history writing focused on humans and anchored in their everyday sense of time.

I will return to Weisman's experiment in the last part of this chapter. There is much in the debate on climate change that should be of interest to those involved in contemporary discussions about history. For as the idea gains ground that the grave environmental risks of global warming have to do with excessive accumulation in the atmosphere of greenhouse gases produced mainly through the burning of fossil fuel and the industrialized use of animal stock by human beings, certain scientific propositions have come into circulation in the public domain that have profound, even transformative, implications for how we think about human history or about what the late C. A. Bayly once called "the birth of the modern world."[2] Indeed, what scientists have said about climate change challenges not only the ideas about the human that usually sustain the discipline of history but also the analytic strategies that postcolonial and postimperial historians have deployed in the last two decades in response to the postwar scenario of decolonization and globalization.

In what follows I present some responses to the contemporary crisis from a historian's point of view. However, a word about my own relationship to the literature on climate change—and indeed to the crisis itself—may be in order. I am a practicing historian with a strong interest in the nature of history as a form of knowledge, and my relationship to the science of global warming is derived, at some remove, from what scientists and other informed writers have written for the education of the general public (sometimes at the risk of irritating their specialist colleagues because of the necessary simplifications that such writing requires). Scientific studies of global warming are often said to have originated with the discoveries of the Swedish scientist Svante Arrhenius in the 1890s, but self-conscious discussions of global warming in the public realm began in the late 1980s and early 1990s, the same period in which social scientists and humanists began to discuss globalization.[3] However, these discussions ran parallel to each other. While globalization, once recognized, was of immediate interest to humanists and social scientists, global warming, in spite of a good number of books published in the 1990s, did not become a public concern until the 2000s. The reasons are not far to seek. As early as 1988 James Hansen, then director of NASA's Goddard Institute of Space Studies, told a Senate committee about global warming and later remarked to a group of reporters on the same day, "It's time to stop waffling . . . and say that the greenhouse effect is here and is affecting our climate."[4] But governments, beholden to special interests and wary of political costs, would not listen. George H. W. Bush, then the president of the United States, famously quipped that he was going to fight the greenhouse effect with the "White House effect."[5] The situation changed in the 2000s when the warnings became dire, and the signs of the crisis—such as drought in Australia, frequent cyclones and brush fires, crop failures in many parts of the world, melting Himalayan and other mountain glaciers and of polar ice caps, increasing acidity of the seas, and damage to the food chain—became politically and economically inescapable. Added to this were growing concerns, voiced by many, about the rapid destruction of other species and about the global footprint of a human population poised to pass the nine billion mark by 2050.[6]

As the crisis gathered momentum in the last few years, I realized that all my readings in theories of globalization, Marxist analysis of capital, subaltern studies, and postcolonial criticism over the last twenty-five years, while enormously useful in studying globalization, had not really prepared me for making sense of this planetary conjuncture within which humanity finds itself today.[7] The change of mood in globalization analysis may be seen by comparing the late Giovanni Arrighi's masterful history of world capitalism, *The Long Twentieth Century* (1994), with his more recent *Adam Smith in Beijing* (2007), which, among other things, seeks to understand the implications of the economic rise of China. The first book, a long meditation on the chaos internal to capitalist economies, ends with the thought of capitalism burning up humanity "in the horrors (or glories) of the escalating violence that has accompanied the liquidation of the Cold War world order." It is clear that the heat that burns the world in Arrighi's narrative comes from the engine of capitalism and not from global warming. By the time Arrighi comes to write *Adam Smith in Beijing*, however, he is much more concerned with the question of ecological limits to capitalism. That theme provides the concluding note of the book, suggesting the distance that a critic such as Arrighi traveled in the thirteen years that separate the publication of the two books.[8] If, indeed, globalization and global warming are born of overlapping processes, the question is, How do we bring them together in our understanding of the world?

Not being a scientist myself, I also make a fundamental assumption about the science of climate change. I assume the science to be right in its broad outlines. I thus assume that the views expressed particularly in the 2007 Fourth Assessment Report of the Intergovernmental Panel on Climate Change (IPCC) of the United Nations, in the *Stern Review on the Economics of Climate Change*, and in the many books that have been published by scientists and scholars seeking to explain the science of global warming leave me with enough rational ground for accepting, unless the scientific consensus shifts in a major way, that there is a large measure of truth to anthropogenic theories of climate change.[9] For this position, I depend on observations such as the following one reported by Naomi Oreskes, then a historian of science at the University of California, San Diego, now working at Harvard. Upon examining the abstracts of 928 papers on global warming published in specialized peer-reviewed scientific journals between 1993 and 2003, Oreskes found that not a single one sought to refute the "consensus" among scientists "over the reality of human-induced climate change." "Virtually all professional climate scientists," writes Oreskes, "agree on the reality of human-induced climate change, but debate continues on tempo and mode."[10] Indeed, in what I have read so far, I have not seen any reason yet for remaining a global-warming skeptic.[11]

The scientific consensus around the proposition that the present crisis of climate change is man-made forms the basis of what I have to say here. In the interest of clarity and focus, I present my propositions in the form of four theses. The last three theses follow from the first one. I begin with the proposition that anthropogenic explanations of climate change spell the collapse of the age-old humanist distinction—prevalent in the seventeenth century but dominant really in the nineteenth—between natural history and human history and end by returning to the question I opened with: How does the crisis of climate change appeal to our sense of human universals while challenging at the same time our capacity for historical understanding?[12]

Thesis 1: Anthropogenic Explanations of Climate Change Spell the Collapse of the Humanist Distinction between Natural History and Human History

Philosophers and students of history have often displayed a conscious tendency to separate human history—or the story of human affairs, as R. G. Collingwood put it—from natural history, sometimes proceeding even to deny that nature could ever have history quite in the same way humans have it. This practice itself has a long and rich past of which, for reasons of space and personal limitations, I can only provide a very provisional, thumbnail, and somewhat arbitrary sketch.[13]

We could begin with the old Viconian-Hobbesian idea that we, humans, could have proper knowledge of only civil and political institutions because we made them, while nature remains God's work and ultimately inscrutable to man. "The true is identical with the created: *verum ipsum factum*" is how Croce summarized Vico's famous dictum.[14] Vico scholars have sometimes protested that Vico did not make such a drastic separation between the natural and the human sciences as Croce and others read into his writings, but even they admit that such a reading is widespread.[15]

This Viconian understanding was to become a part of the historian's common sense in the nineteenth and twentieth centuries. It made its way into Marx's famous utterance that "men make their own history, but they do not make it just as they please" and into the title of the Marxist archaeologist V. Gordon Childe's well-known book *Man Makes Himself*.[16] Croce seems to have been a major source of this distinction in the second half of the twentieth century through his influence on "the lonely Oxford historicist" Collingwood who, in turn, deeply influenced E. H. Carr's 1961 book *What Is History?*, which is still perhaps one of the best-selling books on the historian's craft.[17] Croce's thoughts, one could say, unbeknown to his legatees and with unforeseeable modifications, have triumphed in our understanding of history in the postcolonial age. Behind Croce and his adaptations of Hegel and hidden in Croce's creative misreading of his predecessors stands the more distant and foundational figure of Vico.[18] The connections here, again, are many and complex. Suffice it to say for now that Croce's 1911 book *La filosofia di Giambattista Vico*, dedicated, significantly, to Wilhelm Windelband, was translated into English in 1913 by none other than Collingwood, who was an admirer, if not a follower, of the Italian master.

Collingwood's own argument for separating natural history from human ones developed its own inflections while still running on broadly Viconian lines as interpreted by Croce. Nature, Collingwood remarked, has no "inside." "In the case of nature, this distinction between the outside and the inside of an event does not arise. The events of nature are mere events, not the acts of agents whose thought the scientist endeavours to trace." Hence, "all history properly so called is the history of human affairs." The historian's job is "to think himself into [an] action, to discern the thought of its agent." A distinction, therefore, has to be made "between historical and non-historical human actions.... So far as man's conduct is determined by what may be called his animal nature, his impulses and appetites, it is non-historical; the process of those activities is a natural process." Thus, says Collingwood, "the historian is not interested in the fact that men eat and sleep and make love and thus satisfy their natural appetites; but he is interested in the social customs which they create by their thought as a framework within which these appetites find satisfaction in ways sanctioned by convention and morality." Only the history of the social construction of the body, not the history of the body as such, can be studied. By splitting the human into the natural and the social or cultural, Collingwood saw no need to bring the two together.[19]

In discussing Croce's 1893 essay "History Subsumed under the Concept of Art," Collingwood wrote, "Croce, by denying [the German idea] that history was a science at all, cut himself at one blow loose from naturalism, and set his face towards an idea of history as something radically different from nature."[20] David Roberts gives a fuller account of the more mature position in Croce. Croce drew on the writings of Ernst Mach and Henri Poincaré to argue that "the concepts of the natural sciences are human constructs elaborated for human purposes." "When we peer into nature," he said, "we find only ourselves." We do not "understand ourselves best as part of the natural world." So, as Roberts puts it, "Croce proclaimed that there is no world but the human world, then took over the central doctrine of Vico that we can know the human world because we have made it." For Croce, then, all material objects were subsumed into human thought. No rocks, for example, existed in themselves. Croce's idealism, Roberts explains, "does not mean that rocks, for example, 'don't exist' without human beings to think them. Apart from human concern and language, they neither exist nor do not exist, since 'exist' is a human concept that has meaning only within a context of human concerns and purposes."[21] Both Croce and Collingwood would thus enfold human history and nature—to the extent that the latter could be said to have history—into purposive human action. What exists beyond that does not "exist" because it does not exist for humans in any meaningful sense.

In the twentieth century, however, other arguments, more sociological or materialist, have existed alongside the Viconian one. They, too, have continued to justify the separation of human from natural history. One influential though perhaps infamous example would be the booklet on the Marxist philosophy of history that Stalin published in 1938, *Dialectical and Historical Materialism*. This is how Stalin put the problem:

> Geographical environment is unquestionably one of the constant and indispensable conditions of development of society and, of course, . . . [it] accelerates or retards its development. But its influence is not the determining influence, inasmuch as the changes and development of society proceed at an incomparably faster rate than the changes and development of geographical environment. In the space of 3000 years three different social systems have been successfully superseded in Europe: the primitive communal system, the slave system and the feudal system. . . . Yet during this period geographical conditions in

Europe have either not changed at all, or have changed so slightly that geography takes no note of them. And that is quite natural. Changes in geographical environment of any importance require millions of years, whereas a few hundred or a couple of thousand years are enough for even very important changes in the system of human society.[22]

For all its dogmatic and formulaic tone, Stalin's passage captures an assumption perhaps common to historians of the mid-twentieth century: man's environment did change but changed so slowly as to make the history of man's relation to his environment almost timeless and thus not a subject of historiography at all. Even when Fernand Braudel rebelled against the state of the discipline of history as he found it in the late 1930s and proclaimed his rebellion later in 1949 through his great book *The Mediterranean*, it was clear that he rebelled mainly against historians who treated the environment simply as a silent and passive backdrop to their historical narratives, something dealt with in the introductory chapter but forgotten thereafter, as if, as Braudel put it, "the flowers did not come back every spring, the flocks of sheep migrate every year, or the ships sail on a real sea that changes with the seasons." In composing *The Mediterranean*, Braudel wanted to write a history in which the seasons—"a history of constant repetition, ever-recurring cycles"—and other recurrences in nature played an active role in molding human actions.[23] The environment, in that sense, had an agentive presence in Braudel's pages, but the idea that nature was mainly repetitive had a long and ancient history in European thought, as Gadamer showed in his discussion of Johann Gustav Droysen.[24] Braudel's position was no doubt a great advance over the kind of nature-as-a-backdrop argument that Stalin developed. But it shared a fundamental assumption, too, with the stance adopted by Stalin: the history of "man's relationship to the environment" was so slow as to be "almost timeless."[25] In today's climatologists' terms, we could say that Stalin and Braudel and others who thought thus did not have available to them the idea, now widespread in the literature on global warming, that the climate, and hence the overall environment, can sometimes reach a tipping point at which this slow and apparently timeless backdrop for human actions transforms itself with a speed that can only spell disaster for human beings.

If Braudel to some degree made a breach in the binary of natural/human history, one could say that the rise of environmental history in the late twentieth century made the breach wider. It could even be argued that environmental historians have sometimes indeed progressed toward producing what could be called natural histories of man. But there is a very important difference between the understanding of the human being that these histories have been based on and the agency of the human now being proposed by scientists writing on climate change. Simply put, environmental history—where it was not straightforwardly cultural, social, or economic history—looked on human beings as biological agents. Alfred Crosby Jr., whose book *The Columbian Exchange* did much to pioneer the "new" environmental histories in the early 1970s, put the point thus in his original preface: "Man is a biological entity before he is a Roman Catholic or a capitalist or anything else."[26] The recent book by Daniel Lord Smail *On Deep History and the Brain* is adventurous in attempting to connect knowledge gained from evolutionary sciences and neurosciences with human histories. Smail's book pursues possible connections between biology and culture—between the history of the human brain and cultural history, in particular—while being always sensitive to the limits of biological reasoning. But it is the history of human biology and not any recent theses about the newly acquired geological agency of humans that concerns Smail.[27]

Scholars writing on the current climate-change crisis are indeed saying something significantly different from what environmental historians have said so far. In unwittingly destroying the artificial but time-honored distinction between natural and human histories, climate scientists posit that the human being has become something much larger than the simple biological agent that he or she always has been. Humans now wield a geological force. As Oreskes puts it:

> To deny that global warming is real is precisely to deny that humans have become geological agents, changing the most basic physical processes of the earth.
>
> For centuries, scientists thought that earth processes were so large and powerful that nothing we could do could change them. This was a basic tenet of geological science: that human chronologies were insignificant compared with the vastness of geological time; that human activities were insignificant compared with the force of geological processes. And once they were. But no more. There are now so many of us cutting down so many trees and burning so many billions of tons of fossil fuels that we have indeed become geological agents. We have changed the chemistry of our atmosphere, causing sea level to rise, ice to melt, and climate to change. There is no reason to think otherwise.[28]

Biological agents and geological agents are two different names with very different consequences. Environmental history, to go by Crosby's masterful survey of the origins and the state of the field in 1995, has much to do with biology and geography but hardly ever imagined human impact on the planet on a geological scale. It was still a vision of man "as a prisoner of climate," as Crosby put it quoting Braudel, and not of man as the maker of it.[29] To call human beings geological agents is to scale up our imagination of the human. Humans are biological agents, both collectively and as individuals. They have always been so. There was no point in human history when humans were not biological agents. But climate scientists' claims about human agency introduce a question of scale. Humans can become a planetary geological agent only historically and collectively, that is, when we have reached numbers and invented technologies that are on a scale large enough to have an impact on the planet itself. To call ourselves a geophysical force is to attribute to us a force on the same scale as that released at other times when there has been a mass extinction of species.[30] We seem to be currently going through that kind of a period. The current "rate in the loss of species diversity," specialists argue, "is similar in intensity to the event around 65 million years ago which wiped out the dinosaurs."[31] Our footprint was not always that large. Humans began to acquire this agency only since the Industrial Revolution, but the process really picked up in the second half of the twentieth century. In that sense, we can say that it is only very recently that the distinction between human and natural histories—much of which had been preserved even in environmental histories that saw the two entities in interaction—has begun to collapse. For it is no longer a question simply of man having an interactive relation with nature. This humans have always had, or at least that is how man has been imagined in a large part of what is generally called the Western tradition.[32] Now it is being claimed that humans are a force of nature in the geological sense. A fundamental assumption of Western (and now universal) political thought has come undone in this crisis.[33]

Thesis 2: The Idea of the Anthropocene, the New Geological Epoch When Humans Exist as a Geological Force, Severely Qualifies Humanist Histories of Modernity/Globalization

How to combine human cultural and historical diversity with human freedom has formed one of the key underlying questions of human histories written of the period from 1750 to the years of present-day globalization. Diversity, as Gadamer pointed out with reference to Leopold von Ranke, was itself a figure of freedom in the historian's imagination of the historical process.[34] *Freedom* has, of course, meant different things at different times, ranging from ideas of human and citizens' rights to those of decolonization and self-rule. Freedom, one could say, is a blanket category for diverse imaginations of human autonomy and sovereignty. Looking from the works of Kant, Hegel, or Marx; nineteenth-century ideas of progress and class struggle; the struggle against slavery; the

Russian and Chinese revolutions; the resistance to Nazism and Fascism; the decolonization movements of the 1950s and 1960s and the revolutions in Cuba and Vietnam; the evolution and explosion of the rights discourse; the fight for civil rights for African Americans, Indigenous peoples, Indian *Dalits*, and other minorities; and the kind of arguments that, say, Amartya Sen put forward in his book *Development as Freedom*, one could say that freedom has been the most important motif of written accounts of human history of these two hundred and fifty years. Of course, as I have already noted, freedom has not always carried the same meaning for everyone. Francis Fukuyama's understanding of freedom would be significantly different from that of Sen. But this semantic capaciousness of the word only speaks to its rhetorical power.

In no discussion of freedom in the period since the Enlightenment was there ever any awareness of the geological agency that human beings were acquiring at the same time as—and through processes closely linked to—their acquisition of freedom. Philosophers of freedom were mainly, and understandably, concerned with how humans would escape the injustice, oppression, inequality, or even uniformity foisted on them by other humans or human-made systems. Geological time and the chronology of human histories remained unrelated. This distance between the two calendars, as we have seen, is what climate scientists now claim has collapsed. The period I have mentioned, from 1750 to now, is also the time when human beings switched from wood and other renewable fuels to large-scale use of fossil fuel—first coal and then oil and gas. The mansion of modern freedoms stands on an ever-expanding foundation of fossil-fuel use. Most of our freedoms so far have been energy intensive. The period of human history usually associated with what we today think of as the institutions of civilization—the beginnings of agriculture, the founding of cities, the rise of the religions we know, the invention of writing—began about twelve thousand years ago as the planet moved from one geological period, the last ice age or the Pleistocene, to the more recent and warmer Holocene. The Holocene is the period we are supposed to be in, but the possibility of anthropogenic climate change has raised the question of its termination. Now that humans—thanks to our numbers, technology, the burning of fossil fuel, and other related activities—have become a geological agent on the planet, some scientists have proposed that we recognize the beginning of a new geological era, one in which humans act as a main determinant of the environment of the planet. The name they have coined for this new geological age is Anthropocene. The proposal was first made by the Nobel Prize–winning chemist Paul J. Crutzen and his collaborator, a marine science specialist, Eugene F. Stoermer. In a short statement published in 2000, they said,

"Considering ... [the] major and still growing impacts of human activities on earth and atmosphere, and at all, including global, scales, it seems to us more than appropriate to emphasize the central role of mankind in geology and ecology by proposing to use the term 'anthropocene' for the current geological epoch."[35] Crutzen elaborated on the proposal in a short piece published in *Nature* in 2002:

> For the past three centuries, the effects of humans on the global environment have escalated. Because of these anthropogenic emissions of carbon dioxide, global climate may depart significantly from natural behaviour for many millennia to come. It seems appropriate to assign the term "Anthropocene" to the present ... human-dominated, geological epoch, supplementing the Holocene—the warm period of the past 10–12 millennia. The Anthropocene could be said to have started in the latter part of the eighteenth century, when analyses of air trapped in polar ice showed the beginning of growing global concentrations of carbon dioxide and methane. This date also happens to coincide with James Watt's design of the steam engine in 1784.[36]

It is, of course, true that Crutzen's saying so does not make the Anthropocene an officially accepted geologic period. As Mike Davis comments, "in geology, as in biology or history, periodization is a complex, controversial art" involving, always, vigorous debates and contestation.[37] The name Holocene for "the post-glacial geological epoch of the past ten to twelve thousand years,"[38] for example, gained no immediate acceptance when proposed—apparently by Sir Charles Lyell—in 1833. The International Geological Congress officially adopted the name at their meeting in Bologna after about fifty years in 1885.[39] The same goes for Anthropocene. Scientists have engaged Crutzen and his colleagues on the question of when exactly the Anthropocene may have begun. But the February 2008 newsletter of the Geological Society of America, *GSA Today*, opens with a statement signed by the members of the Stratigraphy Commission of the Geological Society of London accepting Crutzen's definition and dating of the Anthropocene.[40] Adopting a "conservative" approach, they conclude, "Sufficient evidence has emerged of stratigraphically significant change (both elapsed and imminent) for recognition of the Anthropocene—currently a vivid yet informal metaphor of global environmental change—as a new geological epoch to be considered for formalization by international discussion."[41] As this book itself is evidence, the term has now acquired a vigorously contested life in the humanities as well.[42]

So, has the period from 1750 to now been one of freedom or that of the Anthropocene? Is the Anthropocene a critique of the narratives of freedom? Is the geological agency of humans the price we pay for the pursuit of freedom? In some ways, yes. As Edward O. Wilson said in his *The Future of Life*, "Humanity has so far played the role of planetary killer, concerned only with its own short-term survival. We have cut much of the heart out of biodiversity. . . . If Emi, the Sumatran rhino could speak, she might tell us that the twenty-first century is thus far no exception."[43] But the relation between Enlightenment themes of freedom and the collapsing of human and geological chronologies seems more complicated and contradictory than a simple binary would allow. It is true that human beings have tumbled into being a geological agent through their own decisions.[44] The Anthropocene, one might say, has been an unintended consequence of human choices—"unintended" at least for the period when the science of global warming was not generally known, though this does not absolve corporations such as Exxon for developing technologies for extracting "unconventional" oil even after becoming aware of the danger of global warming.[45] But it is also clear that for humans, any thought of the way out of our current predicament cannot but refer to the idea of deploying reason in global, collective public life. As Wilson put it, "We know more about the problem now. . . . We know what to do."[46] Or, to quote Crutzen and Stoermer again,

> Mankind will remain a major geological force for many millennia, maybe millions of years, to come. To develop a world-wide accepted strategy leading to sustainability of ecosystems against human-induced stresses will be one of the great future tasks of mankind, requiring intensive research efforts and wise application of knowledge thus acquired. . . . An exciting, but also difficult and daunting task lies ahead of the global research and engineering community to guide mankind towards global, sustainable, environmental management.[47]

Logically, then, in the era of the Anthropocene, we need the Enlightenment (i.e., reason) even more than in the past. There is one consideration though that must qualify this optimism about the role of reason and that has to do with the most common shape that freedom takes in human societies: politics. Politics has never been based on reason alone. And politics in the age of the masses and in a world already complicated by sharp inequalities between and inside nations is something no one can control. "Sheer demographic momentum," writes Davis, "will increase the world's urban population by 3 billion people over the next 40 years (90% of them in poor cities), and no one—absolutely no one [including, one might say, scholars on the Left]—has a clue how a planet of slums, with growing food and energy crises, will accommodate their biological survival, much less their inevitable aspirations to basic happiness and dignity."[48]

It is not surprising then that the crisis of climate change should produce anxieties precisely around futures that we cannot visualize. Scientists' hope that reason will guide us out of the present predicament is reminiscent of the social opposition between the myth of science and the actual politics of the sciences that Bruno Latour discusses in his *Politics of Nature*.[49] Bereft of any sense of politics, Wilson can only articulate his sense of practicality as a philosopher's hope mixed with anxiety: "Perhaps we will act in time."[50] Yet the very science of global warming produces of necessity political imperatives. Tim Flannery's book, for instance, raises the dark prospects of an "Orwellian nightmare" in a chapter titled "2084: The Carbon Dictatorship?"[51] Mark Maslin concludes his book with some gloomy thoughts: "It is unlikely that global politics will solve global warming. Technofixes are dangerous or cause problems as bad as the ones they are aimed at fixing. . . . [Global warming] requires nations and regions to plan for the next 50 years, something that most societies are unable to do because of the very short-term nature of politics." His recommendation, "we must prepare for the worst and adapt," coupled with Davis's observations about the coming "planet of slums," places the question of human freedom under the cloud of the Anthropocene.[52]

Thesis 3: The Geological Hypothesis Regarding the Anthropocene Requires Us to Put Global Histories of Capital in Conversation with the Species History of Humans

Analytic frameworks engaging questions of freedom by way of critiques of capitalist globalization have not in any way become obsolete in the age of climate change. If anything, as Davis shows, climate change may well end up accentuating all the inequities of the capitalist world order if the interests of the poor and vulnerable are neglected.[53] Capitalist globalization exists; so should its critiques. But these critiques do not give us an adequate hold on human history once we accept that the crisis of climate change is here with us and may exist as part of this planet for much longer than capitalism or long after capitalism has undergone many more historic mutations. The problematic of globalization allows us to read climate change only as a crisis of capitalist management. While there is no denying that climate change has profoundly to do with the history of capital, a critique that is only a critique of capital is not sufficient for addressing questions relating to human history once the crisis of climate change has been acknowledged and the Anthropocene has begun to loom on the horizon of our present. The geologic now of the Anthropocene has become entangled with the now of human history.

Scholars who study human beings in relation to the crisis of climate change and other ecological problems emerging on a world scale make a distinction between the recorded history of human beings and their deep history. Recorded history refers, very broadly, to the eleven thousand or so years that have passed since the invention of agriculture but more usually to the last four thousand years or so for which written records exist. Historians of modernity and "early modernity" usually move in the archives of the last five hundred years. The history of humans that goes beyond these years of written records constitutes what other students of human pasts—not professional historians—call prehistory, and beyond that, deep history. As Wilson, one of the main proponents of this distinction, writes, "Human behavior is seen as the product not just of recorded history, ten thousand years recent, but of deep history, the combined genetic and cultural changes that created humanity over hundreds of [thousands of] years."[54] It of course goes to the credit of Smail that he has attempted to explain to professional historians the intellectual appeal of deep history.[55]

Without such knowledge of the deep history of humanity it would be difficult to arrive at a secular understanding of why climate change constitutes a crisis for humans. Geologists and climate scientists may explain why the current phase of global warming—as distinct from the warming of the planet that has happened before—is anthropogenic in nature, but the ensuing crisis for humans is not understandable unless one works out the consequences of that warming. The consequences make sense only if we think of humans as a form of life and look on human history as part of the history of life on this planet. For ultimately what the warming of the planet threatens is not the geological planet itself but the very conditions, both biological and geological, on which the survival of human species as well as of other forms of life depends. The widely acknowledged threat that the present crisis of biodiversity may indeed balloon into a sixth great extinction of species in the history of the planet constitutes an event horizon for several mainstream narratives of planetary climate change.

The word that scholars such as Wilson or Crutzen use to designate life in the human form—and in other living forms—is *species*. They speak of the human being as a species and find that category useful in thinking about the nature of the current crisis. It is a word that will never occur in any standard history or political-economic analysis of globalization by scholars on the Left, for the analysis of globalization refers, for good reasons, only to the recent and recorded history of humans. Species thinking, on the other hand, is connected to the enterprise of deep history. Further, Wilson and Crutzen actually find such thinking essential to visualizing human well-being. As Wilson writes, "We need this longer view . . . not only to understand our species but more firmly to secure its future."[56] The task of placing, historically, the crisis of climate change thus requires us to bring together intellectual formations that are somewhat in tension with each other: the planetary and the global, deep and recorded histories; species thinking and critiques of capital.

In saying this, I work somewhat against the grain of historians' thinking on globalization and world history. In a landmark essay published in 1995 and titled "World History in a Global Age," Michael Geyer and Charles Bright wrote, "At the end of the twentieth century, we encounter, not a universalizing and single modernity but an integrated world of multiple and multiplying modernities." "As far as world history is concerned," they said, "there is no universalizing spirit. . . . There are, instead, many very specific, very material and pragmatic practices that await critical reflection and historical study." Yet thanks to global connections forged by trade, empires, and capitalism, "we confront a startling new condition: humanity, which has been the subject of world history for many centuries and civilizations, has now come into the purview of all human beings. This humanity is extremely polarized into rich and poor."[57] This humanity, Geyer and Bright imply in the spirit of the philosophies of difference, is not one. It does not, they write, "form a single homogenous civilization." "Neither is this humanity any longer a mere species or a natural condition. For the first time," they say, with some existentialist flourish, "we as human beings collectively constitute ourselves and, hence, are responsible for ourselves."[58] Clearly, the scientists who advocate the idea of the Anthropocene are saying something quite the contrary. They argue that because humans constitute a particular kind of species they can, in the process of dominating other species, acquire the status of a geologic force. Humans, in other words, have become a natural condition, at least today. How do we create a conversation between these two positions?

It is understandable that the biological-sounding talk of species should worry historians. They feel concerned about their finely honed sense of contingency, difference, and freedom in human affairs having to cede ground to a more deterministic view of the world. Besides, there are always, as Smail recognizes, dangerous historical examples of the political use of biology.⁵⁹ The idea of species, it is feared, in addition, may introduce a powerful degree of essentialism in our understanding of humans. I will return to the question of contingency later in this section, but on the issue of essentialism, Smail helpfully points out why species cannot be thought of in essentialist terms:

> Species, according to Darwin, are not fixed entities with natural essences imbued in them by the Creator. . . . Natural selection does not homogenize the individuals of a species [as otherwise natural selection would not work]. . . . Given this state of affairs, the search for a normal . . . nature and body type [of any particular species] is futile. And so it goes for the equally futile quest to identify "human nature." Here, as in so many areas, biology and cultural studies are fundamentally congruent.⁶⁰

It is clear that different academic disciplines position their practitioners differently with regard to the question of how to view the human being. All disciplines have to create their objects of study. If medicine or biology reduces the human to a certain specific understanding of him or her, humanist historians often do not realize that the protagonists of their stories—persons—are reductions too. Absent personhood, there is no human subject of history. That is why Derrida earned the wrath of Foucault by pointing out that any desire to enable or allow madness itself to speak in a history of madness would be "the maddest aspect" of the project.⁶¹ An object of critical importance to humanists of all traditions, personhood is nevertheless no less of a reduction of or an abstraction from the embodied and whole human being than, say, the human skeleton discussed in an anatomy class.

The crisis of climate change calls on academics to rise above their disciplinary prejudices, for it is a crisis of many dimensions. In that context, it is interesting to observe the role that the category of species has begun to play among scholars, including economists, who have already gone further than historians in investigating and explaining the nature of this crisis. The economist Jeffrey Sachs's book *Common Wealth*, meant for the educated but lay public, uses the idea of species as central to its argument and devotes a whole chapter to the Anthropocene.⁶² In fact, the scholar from whom Sachs solicited a foreword for his book was none other than Edward Wilson. The concept of species plays a quasi-Hegelian role in Wilson's foreword in the same way as the multitude or the masses in Marxist writings. If Marxists of various hues have at different times thought that the good of humanity lay in the prospect of the oppressed or the multitude realizing their own global unity through a process of coming into self-consciousness, Wilson pins his hope on the unity possible through our collective self-recognition as a species: "Humanity has consumed or transformed enough of Earth's irreplaceable resources to be in better shape than ever before. We are smart enough and now, one hopes, well informed enough to achieve self-understanding as a unified species. . . . We will be wise to look on ourselves as a species."[63]

Yet doubts linger about the use of the idea of species in the context of climate change, and it would be good to deal with one that can easily arise among critics on the Left. One could object, for instance, that all the anthropogenic factors contributing to global warming—the burning of fossil fuel, the industrialization of animal stock, the clearing of tropical and other forests, and so on—are after all part of a larger story: the unfolding of capitalism in the West and the imperial or quasi-imperial domination by the West of the rest of the world. It is from that recent history of the West that the elite of China, Japan, India, Russia, and Brazil have drawn inspiration in attempting to develop their own trajectories toward superpower politics and global domination through capitalist economic, technological, and military might. If this is broadly true, then does not the talk of species or mankind simply serve to hide the reality of capitalist production and the logic of imperial—formal, informal, or machinic in a Deleuzian sense—domination that it fosters? Why should one include the poor of the world—whose carbon footprint is small anyway—by use of such all-inclusive terms as *species* or *mankind* when the blame for the current crisis should be squarely laid at the door of the rich nations in the first place and of the richer classes in the poorer ones?

We need to stay with this question a little longer; otherwise the difference between the present historiography of globalization and the historiography demanded by anthropogenic theories of climate change will not be clear to us. Though some scientists would want to date the Anthropocene from the time agriculture was invented and some from even earlier—from hominin control of fire, for instance—my readings mostly suggest that our falling into the current phase of the Anthropocene (when we begin to regard ourselves consciously as a geological agent) was neither an ancient nor an inevitable happening. Human civilization surely did not begin on condition that, one day in his history, man would have to shift from wood to coal and from coal to petroleum and gas. That there was much historical contingency in the transition from wood to coal as the main source of energy has been demonstrated powerfully by Kenneth Pomeranz in his pathbreaking book *The Great*

Divergence.⁶⁴ Coincidences and historical accidents similarly litter the stories of the "discovery" of oil, of the oil tycoons, and of the automobile industry as they do any other histories.⁶⁵ Capitalist societies themselves have not remained the same since the beginning of capitalism.⁶⁶ Human population, too, has dramatically increased since the Second World War. India alone is now more than four times more populous than at independence in 1947. Clearly, nobody is in a position to claim that there is something inherent to the human species that has pushed us finally into the Anthropocene. We have stumbled into it. The way to it was no doubt through industrial civilization. (I do not make a distinction here between the capitalist and socialist societies we have had so far, for there was never any principled difference in their use of fossil fuel.)

If the industrial way of life was what got us into this crisis, then the question is, Why think in terms of species, surely a category that belongs to a much longer history? Why could not the narrative of capitalism—and hence its critique—be sufficient as a framework for interrogating the history of climate change and understanding its consequences? It seems true that the crisis of climate change has been necessitated by the high-energy-consuming models of society that capitalist industrialization has created and promoted, but the current crisis has brought into view certain other conditions for the existence of life in the human form that have no intrinsic connection to the logics of capitalist, nationalist, or socialist identities. They are connected rather to the history of life on this planet, the way different life-forms connect to one another, and the way the mass extinction of one species could spell danger for another. Without such a history of life, the crisis of climate change has no human "meaning." For, as I have said before, it is not a crisis for the inorganic planet in any meaningful sense.

In other words, the industrial way of life has acted much like the rabbit hole in Alice's story; we have slid into a state of things that forces on us a recognition of some of the parametric (i.e., boundary) conditions for the existence of institutions central to our idea of modernity and the meanings we derive from them. Let me explain. Take the case of the agricultural revolution, so called, of around 11,700 years ago. It was not just an expression of human inventiveness. It was made possible by certain changes in the amount of carbon dioxide in the atmosphere, a certain stability of the climate, and a degree of warming of the planet that followed the end of the Ice Age (the Pleistocene era)—things over which human beings had no control. "There can be little doubt," writes one of the editors of *Humans at the End of the Ice Age*, "that the basic phenomenon—the waning of the Ice Age—was the result of the Milankovitch phenomena: the orbital and tilt relationships between the Earth and the Sun."[67] The temperature of the planet stabilized within a zone that allowed certain kinds of grass to flourish. Barley and wheat are among the oldest of such grasses. Without this lucky "long summer," or what one climate scientist has called an "extraordinary" "fluke" of nature in the history of the planet, our industrial-agricultural way of life would not have been possible.[68] In other words, whatever our socioeconomic and technological choices, whatever the rights we wish to celebrate as our freedom, we cannot afford to destabilize conditions (such as the temperature zone in which mammalian or plant life survives) that work like boundary parameters of human existence. These parameters are independent of capitalism or socialism. They have been stable for much longer than the histories of these institutions and have allowed human beings to become the dominant species on earth. Unfortunately, we have now ourselves become a geological agent disturbing these parametric conditions needed for our own existence.[69]

This is not to deny the historical role that the richer and mainly Western nations of the world have played in emitting greenhouse gases. To speak of species thinking is not to resist the politics of "common but differentiated responsibility" that China, India, and other developing countries seem keen to pursue when it comes to reducing greenhouse gas emissions.[70] Whether we blame climate change on those who are retrospectively guilty—that is, blame the West for its past performance—or those who are prospectively guilty—China has just surpassed the United States as the largest emitter of carbon dioxide, though not on a per capita basis—is a question that is tied no doubt to the histories of capitalism and modernization.[71] But scientists' discovery of the fact that human beings have in the process become a geological agent points to a shared catastrophe that we have all fallen into. Here is how Crutzen and Stoermer describe that catastrophe:

> The expansion of mankind . . . has been astounding. . . . During the past 3 centuries human population increased tenfold to 6000 million, accompanied e.g. by a growth in cattle population to 1400 million (about one cow per average size family). . . . In a few generations mankind is exhausting the fossil fuels that were generated over several hundred million years. The release of SO_2 . . . to the atmosphere by coal and oil burning, is at least two times larger than the sum of all natural emissions . . . ; more than half of all accessible fresh water is used by mankind; human activity has increased the species extinction rate by thou-

sand to ten thousand fold in the tropical rain forests.... Furthermore, mankind releases many toxic substances in the environment.... The effects documented include modification of the geochemical cycle in large freshwater systems and occur in systems remote from primary sources.[72]

Explaining this catastrophe calls for a conversation between disciplines and between recorded and deep histories of human beings in the same way that the agricultural revolution of twelve thousand years ago could not be explained except through a convergence of three disciplines: geology, archaeology, and history.[73]

Scientists such as Wilson or Crutzen may be politically naive in not recognizing that reason may not be all that guides us in our effective collective choices—in other words, we may collectively end up making some unreasonable choices—but I find it interesting and symptomatic that they speak the language of the Enlightenment. They are not necessarily anticapitalist scholars, and yet clearly they are not for business-as-usual capitalism either. They see knowledge and reason providing humans not only a way out of this present crisis but a way of keeping us out of harm's way in the future. Wilson, for example, speaks of devising a "wiser use of resources" in a manner that sounds distinctly Kantian.[74] But the knowledge in question is the knowledge of humans as a species, a species dependent on other species for its own existence, a part of the general history of life. Changing the climate—increasingly not only the average temperature of the planet but also the acidity and the level of the oceans—and destroying the food chain are actions that cannot be in the interest of our lives. Biodiversity is important for human flourishing irrespective of our political choices. It is therefore impossible to understand global warming as a crisis without engaging the propositions put forward by these scientists. At the same time, the story of capital, the contingent history of our falling into the Anthropocene, cannot be denied by recourse to the idea of species, for the Anthropocene would not have been possible, even as a theory, without the history of industrialization. How do we hold the two together as we think the history of the world since the Enlightenment? How do we relate to a universal history of life—to universal thought, that is—while retaining what is of obvious value in our postcolonial suspicion of the universal? The crisis of climate change calls for thinking simultaneously on both registers, to mix together the immiscible chronologies of capital and species history. This combination, however, stretches, in quite fundamental ways, the very idea of historical understanding.

Thesis 4: The Crosshatching of Species History and the History of Capital Is a Process of Probing the Limits of Historical Understanding

Historical understanding, one could say following the Diltheyan tradition, entails critical thinking that makes an appeal to some generic ideas about human experience. As Gadamer pointed out, Dilthey saw "the individual's private world of experience as the starting point for an expansion that, in a living transposition, fills out the narrowness and fortuitousness of his private experience with the infinity of what is available by re-experiencing the historical world." "Historical consciousness" in this tradition is thus "a mode of self-knowledge" garnered through critical reflections on one's own and others' (historical actors') experiences.[75] Humanist histories of capitalism will always admit of something called the experience of capitalism. E. P. Thompson's brilliant attempt to reconstruct working-class experience of capitalist labor, for instance, does not make sense without that assumption.[76] Humanist histories are histories that produce meaning through an appeal to our capacity not only to reconstruct but, as Collingwood would have said, to reenact in our own minds the experience of the past.

When Wilson then recommends in the interest of our collective future that we achieve self-understanding as a species, the statement does not correspond to any historical way of understanding and connecting pasts with futures through the assumption of there being an element of continuity to human experience. (See Gadamer's point mentioned above.) Who is the we? We humans never experience ourselves as a species. We can only intellectually comprehend or infer the existence of the human species but never experience it as such. There could be no phenomenology of us as a species. Even if we were to identify emotionally with a word like *mankind*, we would not know what being a species is, for in species history, humans are only an instance of the concept species as indeed would be any other life-form. But one never experiences being a concept. The concept dog, Althusser once famously said, drawing on Spinoza, does not bark![77]

I may here, in parenthesis, mention a thoughtful objection that was raised by Ursula Heise against my statement "one never experiences being a concept [species]" after this chapter was published in its first version as an essay. "Granted," she wrote, "humans may not normally be able to experience themselves as a species—any more than they are able to experience themselves as a nation: unless, that is, communities produce institutions, laws, symbols, and forms of rhetoric that establish such abstract categories as perceptible and livable frameworks of experience."⁷⁸ It is no doubt true, as Derrida and others have pointed out, that we have an everyday sense of being individual members of the "human" species through what we precisely share and do not share with other animals around us.⁷⁹ But when I speak of humans constituting a certain formation of domination—a complex of humans, their technologies, and the animal species that flourish through their association with humans—I speak of a certain dominant collectivity that even contains the nonliving (i.e., technology) as part of itself.⁸⁰ This collectivity, cognitively available to me, is still not available to my phenomenological experience of the world. Heise is right: abstract categories like "nation" and "labor" enter our everyday life precisely because there are institutions organized around these categories, such as the United Nations or the Secretary of Labor or trade unions. If the earth's history had reached a point where we had a multispecies organization of governance—something like, say, a Latourian world parliament or a United Organization for Multi-Species Governance—that allowed polar bears, for example, to voice their complaints against humans and ask for adjudication, the category "dominant species" could indeed be part of what Heise calls "lived, existential [and political] relations" and carry a meaning in our everyday experience.⁸¹ But that is still a far cry.

The discussion about the crisis of climate change can then—given the planetary and experience-distant nature of human agency as a geophysical force—produce affect and knowledge about collective human pasts and futures that work at the limits of historical understanding. We experience specific effects of the crisis but not the whole phenomenon. This is often the problem of communicating the science of climate change to local communities on the ground—the specific impacts are concrete and experienceable, while the science is too abstract and planetary. Do we then say, with Geyer and Bright, that "humanity no longer comes into being through 'thought,'"⁸² or do we say with Foucault that "the human being no longer has any history"?⁸³ Geyer and Bright go on to write in a Foucauldian spirit: "Its [world history's] task is to make transparent the lineaments of power, underpinned by information, that compress humanity into a single humankind."⁸⁴

This critique that sees humanity as an effect of power is, of course, valuable for all the hermeneutics of suspicion that it has taught postcolonial scholarship. It is an effective critical tool in dealing with national and global formations of domination. But I do not find it adequate in dealing with the crisis of global warming. First, inchoate figures of us all and other imaginings of humanity invariably haunt our sense of the current crisis. How else would one understand the title of Weisman's book, *The World without Us*, or the appeal of his brilliant though impossible attempt to depict the experience of New York after "we" are gone![85] Second, the wall between human and natural history has been breached. We may not experience ourselves as a geological agent, but we appear to have become one at the level of our being a species, our possession of global technology, and our domination of life on the planet. And without that knowledge that defies historical understanding (in the phenomenological sense explained above), there is no making sense of the current crisis that affects us all. Climate change, refracted through global capital, will no doubt accentuate the logic of inequality that runs through the rule of capital; some people will no doubt gain temporarily at the expense of others. But the whole crisis cannot be reduced to a story of capitalism. Unlike in the crises of capitalism, there are no lifeboats here for the rich and the privileged (witness the frequent bushfires in Australia or recent fires in the wealthy neighborhoods of California).[86] The fires are revisiting both places as I write this sentence in December 2019.

The anxiety global warming gives rise to is reminiscent of the days when many feared a global nuclear war. But there is a very important difference. A nuclear war would have been a conscious decision on the part of the powers that be. Climate change has largely been a combination of intended and unintended consequence of a cascade of human decisions and actions, and it shows, only through scientific analysis, the long-term planetary effects of our actions as a species. While scientific accounts of evolution, mass extinctions, and natural selection would not work without categories like species and speciation, the category "species" has long been recognized to be haunted by philosophical problems of what David N. Stamos called "realism," "conceptualism," and "nominalism."[87] In my argument, *species* may indeed be the name of a placeholder for an emergent, new universal history of humans that flashes up in the moment of the danger that is climate change. But we can never *understand* (in the Diltheyan sense) this universal. It is not a Hegelian universal arising dialectically out of the movement of history or a universal of capital brought forth by the present crisis. Geyer and Bright are right to reject those two varieties of the universal. Yet climate change poses for us a question of a human collectivity, an us pointing to a figure of the universal that escapes our capacity to experience the world. It is more like a universal that arises from a shared sense of a catastrophe. It calls for a global approach to politics without the myth of a global identity, for, unlike a Hegelian universal, it cannot subsume particularities. Borrowing from Adorno, we may provisionally call it a "negative universal history."[88]

Addendum: A Note on Species and Negative Universal History

The observations with which I concluded the first version of this chapter elicited an interesting and sharp comment from Ursula Heise. It is worth engaging with her criticisms because the discussion will, I hope, illuminate the larger argument I am trying to make in this book. Heise wrote,

> Chakrabarty's rejection of species as a concept that might ground collective identity resonates with Dale Jamieson's rejection of species as a relevant category in the interaction with nonhumans, which I quoted and criticized. . . . Chakrabarty's skepticism toward species thinking leaves his argument, which is essentially a call for what in other theoretical discourses would be referred to as a kind of cosmopolitanism, with no positive content. What he imagines at the end is a "negative universalism" that cannot take on a concrete content that would always be less than universal, in that it would be bound to postulate some characteristics of a particular community as the paradigm by which other communities should be measured.[89]

Heise is right to observe that the "negative universal" I try to invoke has no "concrete" positive content. It is empty in that it is an emergent concept with no particular, concrete content yet. But then here is a problem. When we think of the climate crisis as a problem to be solved in historical time, we think of solutions that, theoretically, affect if not embrace humans and nonhumans in that all imagined solutions assume some stable and sustainable relationships between humans and nonhumans (including the nonliving, such as the earth). This is an ambition toward what Heise rightly recognizes as a new form of cosmopolitanism.

One can observe this legitimate ambition in other commentators on the current crisis as well. Jason Moore, for instance, begins his *Capitalism in the Web of Life* with almost a mystical quest for "the politics of liberation for *all* life."[90] But the ambition of Moore's imagination is clear. It is reaching out toward an "all" that is more than human. Similarly, in their book *The Ecological Rift*, Marxist ecologists John Bellamy Foster, Brett Clark, and Richard York describe their vision of a sustainable development that requires replacing "the capitalist system" with an equally mystical "new human whole" that would help maintain "the conditions of life for the millions of other species on Earth."[91]

How do we imagine the totality of this "we" that is larger than human? The reason why someone like Adorno had to think about "negative universal history" in considering issues of history and freedom was that he knew that positing any positive content for "all" of humanity would in fact lead to one particular section of humanity oppressing another particular section in the name of the universal or the whole. In such a situation, Adorno argued, both that which claims to stand for a totality and that which claims to represent difference—the nonidentical that may take the "form of what are more or less natural categories" while being "merely relics from older historical epochs"—"go rancid and become poisonous." "They go rancid," wrote Adorno, "much as the universal principle does when confronted with them." Adorno's example was the civil war in (formerly) Belgian Congo (Zaire from 1971) in the mid-1960s in which Belgian troops were involved. Adorno thought that one could test his thesis "against the recent events in Africa—if indeed we can pluck up the courage to do so, something that is not altogether easy."

> It is really the case that, under the rule of totality, even the particular that opposes it nevertheless collaborates in weaving the web of disaster. It does so not just by lapsing into particularity, but by degenerating into something poisonous and bad. That is to say, these natives who are running wild in Africa for the last time are not one whit better . . . than the barbaric paratroopers who are struggling to make them see reason, i.e., to accept the benefits of a progressive civilization. . . . This great historical trend sucks the marrow out of everything oppositional and recalcitrant.[92]

A "negative universal history" is therefore one that allows the particular to express its resistance to its imbrication in the totality without denying being so imbricated.

Harriet Johnson concludes her study of Adorno's idea of "negative universal history" by saying, "The Anthropocene challenges us to decipher a new universal history because we encounter a set of planetary forces and temporal scales that could not be a direct object of experience in our lives yet will be a determining factor for them. Adorno is important because he looked for ways to tell such stories without, in turn, naturalizing the extant power relations of social history."[93] Imagine taking this proposition beyond the province of human history in which it originates in Johnson's essay. A "negative universal history" in the age of the Anthropocene cannot simply be about humans alone. At the same time, it cannot be about a totality, for then it would simply reproduce all the problems that led Adorno to formulate his propositions around the figure of the negative. Just as in human history, here too, that which is nonidentical to totality has to be able to express itself through resisting its complete incorporation into the totality even as it is so incorporated—thus the project of "provincializing Europe." Similarly, in the case of the "negative universal history" of the Anthropocene, the nonhuman should be able to make itself heard without having to be anthropomorphized or without having to speak the language of humans.[94]

*   *   *

We are not yet at a point in global history where such a prospect seems practical, though one may go to the histories of Indigenous peoples to learn some exemplary lessons on some of the principles involved here.[95] The "negative universal history" of the Anthropocene—the history that gestures to a "we" that may indeed be more than human—can only be an ethical advisory at this point. Its empirical content for now remains necessarily empty. For an "ought" position does not dictate the actual working out of history, though it can give us a supervening perspective—something like Karl Jasper's "epochal consciousness"—on our contemporary debates without prejudging or preempting them.[96] It may someday be possible to fill out the "we" of a negative universal history of the Anthropocene with concrete identities of humans and nonhumans. Or it may not.

* 2 *
Conjoined Histories

As I argued in the last chapter, Anthropogenic global warming brings into view the collision—or the running up against one another—of three histories that from the point of view of human history are normally assumed to be working at such different and distinct paces that they are treated as processes separate from one another for all practical purposes: the history of the Earth system, the history of life including that of human evolution on the planet, and the more recent history of industrial civilization (for many, capitalism). Humans now unintentionally straddle these three histories, which operate on different scales and at different speeds.

The everyday language with which we speak of the climate crisis is shot through with this problem of human and unhuman scales of time. Take the most ubiquitous distinction we make in our everyday prose between nonrenewable sources of energy and the "renewables." We consider fossil fuels nonrenewable on our terms, but as Bryan Lovell—a geologist who worked as an advisor for British Petroleum and an ex-president of the Geological Society of London—points out, fossil fuels are renewable if only we think of them on a scale that is (in his terms) *inhuman*: "Two hundred million years from now, a form of life requiring abundant oil for some purpose should find that plenty has formed since our own times."[1] Indeed, one way to think about the current crisis of anthropogenic climate change is to think of it as a problem of mismatched temporalities. Human institutions and practices are geared to a human sense of time and history. But we now have to use these institutions to address processes that unfold over much larger scales of time.

Paleoclimatologists, for instance, tell a very long history when it comes to explaining the significance of anthropogenic global warming. There is, first of all, the question of evidence. Ice-core samples of ancient air—more than 800,000 years old—have been critical in establishing the anthropogenic nature of the current warming.[2] There are, besides, paleoclimatic records of the past in fossils and other geological materials. In his lucid book on the oil industry's response to the climate crisis—not always or uniformly negative though there is the Exxon example to the contrary—Lovell writes that the people within the industry who supplied it with compelling evidence of the serious challenge that greenhouse gas emissions posed to the future of humanity were geologists who could read deep climate histories buried in sedimentary rocks to see the effects of "a dramatic warming event that took place 55 million years ago." This event has often been cited to illustrate the effects that warming of the surface temperature of the earth can have on the history of life. It is known as the late Paleocene-Eocene Thermal Maximum (PETM).

> Comparison of the volume of carbon released to the atmosphere [then] . . . and the volume we are now releasing ourselves strongly suggests that we are indeed facing a major global challenge. We are in danger of repeating that 55 million-year-old global warming event, which disrupted Earth over 100,000 years. That event took place long before Homo sapiens was around to light so much as a campfire.[3]

How far the arc of the geological history explaining the present climate crisis projects into the future may be quickly seen from the very subtitle of David Archer's *The Long Thaw: How Humans Are Changing the Next 100,000 Years of Earth's Climate*. "Mankind is becoming a force in climate comparable to the orbital variations that drive glacial cycles," writes Archer.[4] "The long lifetime of fossil fuel CO_2," he continues, "creates a sense of fleeting folly about the use of fossil fuels as an energy source. Our fossil fuel deposits, 100 million years old, could be gone in a few centuries, leaving climate impacts that will last for hundreds of millennia. The lifetime of fossil fuel CO_2 in the atmosphere is a few centuries, plus 25% that lasts essentially forever."[5] The carbon cycle of the earth—as Archer explains and as Curt Stager repeats—will eventually clean up the excess CO_2 we put out in the atmosphere, but it works on an unhumanly long timescale.[6]

The climate crisis thus produces problems that we ponder on very different and often incompatible scales of time. Policy specialists think in terms of years, decades, at most centuries, while politicians in democracies think in terms of their electoral cycles. Understanding what anthropogenic climate change is and how long its effects may last calls for thinking on very large and small scales at once, including scales that defy the usual measures of time that inform human affairs. This is another reason that makes it difficult to develop a comprehensive politics of climate change. Archer goes to the heart of the problem here when he acknowledges that the million-year timescale of the planet's carbon cycle is "irrelevant for political considerations of climate change on human time scales." Yet, he insists, it remains relevant to any understanding of anthropogenic climate change because "ultimately the global warming climate event will last for as long as it takes these slow processes to act."[7]

Significant gaps thus open up in the existing literature on the climate problem between cognition and action, between what we scientifically know about it—the vastness of its unhuman scale, for instance—and how we think about it when we treat it as a problem to be handled by the human means and institutions at our disposal. The latter have been developed for addressing problems we face on familiar scales of time. I call these gaps or openings in the landscape of our thoughts rifts because they are like fault lines on a seemingly continuous surface; we have to keep crossing or straddling them as we think or speak of climate change. They inject a certain degree of contradictoriness in our thinking, for we are being asked to think about different scales at once.

I want to discuss here three such rifts: the various regimes of probability that govern our everyday lives in modern economies and which now have to be supplemented by our knowledge of the radical uncertainty of the climate; the story of our necessarily divided human lives having to be supplemented by the story of our collective life as a species, a dominant species, on the planet; and the necessity of making room within our inevitably anthropocentric thinking for forms of disposition toward the planet that do not put humans first. We have not yet overcome these dilemmas to settle decidedly on any one side of them. They remain as rifts.

In what follows, I elaborate on these rifts with a view to demonstrating that the analytics of capital (or of the market), while necessary in the spheres of policy and politics irrespective of where one stands on the question of capitalism, are insufficient instruments in helping us come to grips with the historical significance of anthropogenic climate change. I will go on to conclude by proposing that the climate crisis makes visible an emergent but critical distinction between categories of the globe and the planet that will need to be explored further in order to develop a perspective on the human meaning(s) of global warming and the Anthropocene. Chapter 3 is devoted to the task of developing this distinction.

Probability and Radical Uncertainty

Modern life is ruled by regimes of probabilistic thinking. From evaluating lives for actuarial ends to the working of money and stock markets, we manage our societies by calculating risks and assigning probability values to them.[8] "Economics," writes Charles S. Pearson, "often makes a distinction between risk, where probabilities of outcomes are known, and uncertainty, where probabilities are not known and perhaps unknowable."[9] This is surely one reason why economics as a discipline has emerged as the major art (or "science," as some would like to think of it) of social management today.[10] There is, therefore, an understandable tendency in both climate-justice and climate-policy literature—the latter dominated by economists or legal scholars who think like economists—to focus not so much on what paleoclimatologists or geophysicists who study planetary climate historically have to say about climate change but rather on what we might call the physics of global warming that often presents a predictable, isolated set of relationships of probability and proportion: if the share of greenhouse gases in the atmosphere goes up by x, then the probability of the earth's average surface temperature going up by so much is y.[11]

Such a way of thinking assumes a kind of stability or predictability—however probabilistic it may be—on the part of a warming atmosphere that paleoclimatologists, focused more on the greater danger of tipping points, often do not assume. This is neither because policy thinkers are not concerned about the dangers of climate change nor because they are ignorant of the profoundly nonlinear nature of the relationship between greenhouse gases and the rise in the planet's average surface temperature. But their methods are such that they appear to hold or bracket climate change as a broadly known variable (converting its uncertainties into risks that have been acknowledged and evaluated) while working out practical options humans can create while striving together or even wrangling among themselves. The world climate system, in other words, has no significant capacity to be a wild card in their calculations insofar as they can make policy prescriptions; it is there in a relatively predictable form to be managed by human ingenuity and political mobilization.[12]

The rhetoric of the climate scientists in what they write to persuade the public, on the other hand, is often remarkably vitalist. In explaining the danger of anthropogenic climate change, they often resort to a language that portrays the climate system as a living organism. There is not only the famous case of James Lovelock comparing life on the planet to a single living organism that he christened Gaia—a point that even the "sober" Archer accommodates in his primer on the global carbon cycle as a fair but "philosophical definition."[13] Archer himself describes the "carbon cycle of the Earth" as "alive."[14] The image of climate as a temperamental animal also inhabits the language of Wallace (Wally) Broecker, who, with the help of Robert Kunzig, thus describes his studies, emphasizing the importance of history as a method in the study of climate:

> Every now and then, . . . nature has decided to give a good swift kick to the climate beast. And the beast has responded, as beasts will — violently and a little unpredictably. Computer models . . . [are] certainly a valid approach. But studying how the beast has responded in the past under stress is another way to prepare ourselves for what might happen as we take a whack at it ourselves. That's the idea that has obsessed Broecker for the past twenty-five years, and with each passing year it has come to seem more urgent.[15]

Or notice how Hansen uses the vitalist image of "lethargy" in explaining climate change:

> The speed of glacial-interglacial change is dictated by 20,000-, 40,000-, and 100,000-year time scales for changes of Earth's orbit — but this does not mean that the climate system is inherently *that* lethargic. On the contrary. Human-made climate forcing, by paleoclimate standards, is large and changes in decades, not tens of thousands of years.[16]

The vitalism of this prose does not arise because climate scientists are less "scientific" than economists and policy makers. It issues from climate scientists' anxiousness to communicate and underscore two points about Earth's climate: that its many uncertainties cannot ever be completely tamed by existing human knowledge, and that its exact tipping points are inherently unknowable. As Archer puts it,

> The IPCC forecast for climate change in the coming century is for a generally smooth increase in temperature. . . . However, actual climate changes in the past have tended to be abrupt. . . . Climate models . . . are for the most part unable to simulate the flip flops in the past climate record very well.[17]

It is in fact this sense of a temperamental "climate beast" that is missing from both the literature inspired by economics and by political commitments on the Left. John Broome, a lead author of the Working Group III of the IPCC 2007 report and himself an economist turned philosopher, looks forward to a future where climate models continue to "narrow" the probabilities that "should be assigned to various possibilities." For economic reasoning to have a better grasp of the world, "detailed information about probabilities" is needed, and, adds Broome, "we are waiting for it to be supplied by scientists."[18] But this may misunderstand the nature of the planet's climate and of the models humans make of it. Climate uncertainties may not always be like measurable risks. "Do we really need to know more than we know now about how much the Earth will warm? *Can* we know more?" asks Paul Edwards rhetorically. "It is now virtually certain that CO_2 concentrations will reach 550 ppm (the doubling point) sometime in the middle of this century," and the planet "will almost certainly overshoot CO_2 doubling." Climate scientists, he reports, are engaged in the speculation "that *we will probably never get a more exact estimate than we already have.*"[19]

The reasoning behind Edwards's statement is relevant to my argument. "If engineers are sociologists," writes Edwards, "then climate scientists are historians." Like historians, "every generation of climate scientists revisit the same data, the same events—digging through the archives to ferret out new evidence, correct some previous interpretation," and so on. And "just as with human history, we will never get a single, unshakable narrative of the global climate's past. Instead we get versions of the atmosphere, ... convergent yet never identical."[20] Moreover, "all of today's analyses are based on the climate we have experienced in historical time." "Once the world has warmed by 4°C," he quotes scientists Myles Allen and David Frame, "conditions will be so different from anything we can observe today (and still more different from the last ice age) that it is inherently hard to say when the warming will stop." Their point, Edwards explains, is this: not only do we not know whether "there is some 'safe' level of greenhouse gases that would 'stabilize' the climate" for humans; thanks to anthropogenic global warming, we may "never" be in a position to find out whether such a point of stabilization can exist in human timescales.[21]

The first rift that I speak of thus organizes itself around the question of the tipping point of the climate, a point beyond which global warming could be catastrophic for humans. That such a possibility exists is not in doubt. Paleoclimatologists know that the planet has undergone such warming in the geological past (as in the case of the PETM event). But we cannot predict how quickly such a point could arrive. It remains an uncertainty that is not amenable to the usual cost-benefit analyses that are a necessary part of risk-management strategies. As Pearson explains, "BC [benefit-cost analysis] is not well suited for making catastrophe policy," and he acknowledges that the "special features that distinguish uncertainty in global warming are the presence of nonlinearities, thresholds and potential tipping points, irreversibility, and the long time horizon" that make "projections of technology, economic structure, preferences and a host of other variables 100 years from now increasingly questionable."[22] "The implication of uncertainty, thresholds, tipping points," he writes, "is that we should take a precautionary approach," that is, "avoid taking steps today that lead to irreversible changes."[23] But "the precautionary principle," as Cass Sunstein explains, also involves cost-benefit analysis and some estimation of probability: "Certainly we should acknowledge that a small probability (say, 1 in 100,000) of serious harm (say, 100,000 deaths) deserves extremely serious attention."[24] But we simply don't know the probability of the tipping point being reached over the next several decades or by 2100, for the tipping point would be a function of the rise in global temperature and multiple, unpredictable amplifying feedback loops working together. Under the circumstances, the one principle that James Hansen recommends to policy thinkers concerns the use of coal as a fuel. He writes, "If we want to solve the climate problem, we must phase out coal emissions. Period."[25] Not quite a "precautionary principle" but what in the literature on risks would be known as "the maximin principle": "choose the policy with the best worst-case outcome,"[26] But this would seem unacceptable to governments and businesses around the world; without coal, on which China and India are still dependent to a large degree (68–70 percent of their energy supply), how would the majority of the world's poor be lifted out of poverty in the next few decades and thus be equipped to adapt to the impact of climate change? Or would the world, scrambling to avoid the tipping point of the climate, make the global economy itself tip over and cause untold human misery? Thus, the question arises, Would avoiding "the harm" itself do more harm, especially as we do not know the probability of reaching the tipping point in the coming few decades? This is the dilemma that goes with the application here of the precautionary or the maximin principle, as both Sunstein and Pearson explain.[27] It is not surprising that Stephen Gardiner's chapter on cost-benefit analyses in the context of climate change is titled "Cost-Benefit Paralysis."[28]

At the heart of this rift is the question of scale. On the much more extended canvas on which they place the history of the planet, paleoclimatologists see climatic tipping points and the accompanying possibility of widespread species extinction—as happened during the PETM—as perfectly repeatable phenomena irrespective of whether or not we can model for them. Our strategies of risk management, however, arise from more human calculations of costs and their probabilities over plausible human timescales. The climate crisis requires us to move back and forth between thinking on these different scales all at once.

Our Divided Lives as Humans and Our Collective Life as a Dominant Species

Human-induced climate change gives rise to large and diverse issues of justice: justice between generations, between small island nations and the polluting countries (both past and prospective), and between developed, industrialized nations (historically responsible for most emissions) and the newly industrializing ones. Peter Newell and Matthew Paterson express a sense of discomfiture about the use of the word *human* in the expression "human-induced climate change." "Behind the cosy language used to describe climate change as a common threat to all humankind," they write, "it is clear that some people and countries contribute to it disproportionately, while others bear the brunt of its effects. What makes it a particularly tricky issue to address," they go on to say, "is that it is the people that will suffer most that currently contribute least to the problem, i.e. the poor in the developing world. Despite often being talked about as a scientific question, climate change is *first and foremost* a deeply political and moral issue."[29] In her endorsement of their book, the Indian environmentalist Sunita Narain remarks that "Climate Change we know is intrinsically linked to the model of economic growth in the world."[30] The climate crisis—write John Bellamy Foster, Brett Clark, and Richard York in their important book *The Ecological Rift*—is "at bottom, the product of a social rift: the domination of human being by human being. The driving force is a society based on class, inequality, and acquisition without end."[31]

A very similar position was put forward in 2009 when the Department of Economic and Social Affairs of the United Nations published a report carrying the title *Promoting Development and Saving the Planet*.[32] In signing the report, Sha Zukang, UN undersecretary general for economic and social affairs, wrote, "The climate crisis *is the result of* the very uneven pattern of economic development that evolved over the past two centuries, which allowed today's rich countries to attain their current levels of income, in part through not having to account for the environmental damage now threatening the lives and livelihoods of others."[33] Characterizing climate change as a "development challenge," Sha went on to remark how a certain deficit of trust marks the attitude of the non-Western countries towards the West.[34] The report expanded on his point: "How developing countries can achieve catch-up growth and economic convergence in a carbon-constrained world and what the advanced countries must do to relieve their concerns have become leading questions for policy makers at the national and international levels."[35] The original formulation of this position, to the best of my knowledge, goes back to 1991 when two well-known and respected Indian environmental activists, the late Anil Agarwal and Sunita Narain, authored a booklet titled *Global Warming in an Unequal World: A Case of Environmental Colonialism*, published by their organization, the Centre for Science and Environment, in Delhi.[36] This booklet did much to generate the idea of *common but differentiated responsibilities* and the tendency to argue from figures of per capita emissions of greenhouse gases that became popular as part of the Kyoto Protocol.[37]

There are good reasons why questions of justice arise. Only a few nations (some twelve or fourteen, including China and India in the last decade or so) and a fragment of humanity (about one-fifth) are historically responsible for most of the emissions of greenhouse gases so far. This is true. But we would not be able to differentiate between humans as actors and the planet itself as an actor in this crisis if we did not realize that, leaving aside the question of intergenerational ethics that concerns the future, anthropogenic climate change is not inherently—that is, logically—a problem of past or accumulated intrahuman injustice.

My point here depends on the validity of a distinction often made between a necessary and logical relationship between two entities and a contingent and historical relationship between the same. Making this distinction allows me to make room within my framework for planetary processes that work regardless of how human societies are internally structured. The surface temperature of the planet depends on the extent of greenhouse gases emitted into the atmosphere. The atmosphere does not care whether the gases come from a massive volcanic eruption or internally unjust human societies. To say this is not to deny the *historical* role played by what we think of as global capitalism. Historically speaking, it is, of course, true that the richer nations are responsible for most of the emissions of greenhouse gases as they pursued models of development that produced an unequal world. But imagine the counterfactual reality of a more evenly prosperous and just world made up of the same number of people as today and based on exploitation of cheap energy sourced from fossil fuel. Such a world would undoubtedly be more egalitarian and just—at least in terms of distribution of income and wealth—but the climate crisis could be worse! Our collective carbon footprint could even be larger than it is today—for the world's poor do not consume much and contribute little to the production of greenhouse gases. The climate crisis could have been on us much sooner and in a much more drastic way. It is, ironically, thanks to the poor—that is, to the fact that development *is* uneven and unfair—that we do not put even larger quantities of greenhouse gases into the atmosphere than we actually do. Thus, logically speaking, the warming crisis is really a matter of the quantity of greenhouses gases we put out and into the atmosphere. Those who connect climate change causally to historical origins/formations of economic inequalities in the modern world raise valid questions about historical inequalities, but seeing that as the only cause not only reduces the problem of climate change to that of capitalism (folded into the histories of modern European expansion and empires), it also blinds us to the action—or agency, if you will—of Earth system processes and their unhuman temporalities. In the end, we lose sight of the nature of our present that is defined by the coming together of the relatively short-term processes of human history and other much longer-term processes that belong to Earth systems history and the history of life on the planet.

Agarwal and Narain's insistence, however, that the natural carbon sinks—such as the oceans—are part of the global commons and hence best distributed among nations by applying the principle of equal access on a per capita basis if the world were to aspire "to such lofty ideals like global justice, equity and sustainability" raises by implication a very important issue: the simultaneously acknowledged and disavowed problem of population.[38] Population is often the elephant in the room in discussions of climate change. Population is a complex question and does not have to raise the bogey of Malthusianism with which it has often been associated in the past, an association that makes any discussion of it difficult to undertake.[39] There is no blanket "population problem." The population question is complex because the question of "overpopulation" is also not simple. One could plausibly argue, for instance, that the developed countries are "overpopulated" if one looked simply at figures of consumption, while wild animals losing their habitat to an ever-expanding poor and rapidly urbanizing population may be a problem characteristic of a place like India. The presently large number of humans on the planet—while due surely in part to modern medicine, public health measures, personal hygiene, eradication of epidemics, the use of artificial fertilizers, and so on—cannot be attributed in any straightforward way to a logic of a predatory and capitalist West, for neither China nor India pursued unbridled capitalism in the decades when their populations exploded. If India had been more successful with population control or with economic development, her per capita emission figures would have been higher (that the richer classes in India want to emulate Western styles and standards of consumption would be obvious to any observer). Indeed, the Indian minister in charge of the environment and forests, Jairam Ramesh, said as much in an address to the Indian parliament in 2009: "per-capita is an accident of history. It so happened that we could not control our population."[40]

Yet population remains a very important factor in how the climate crisis plays out. Chinese and Indian governments continue to build coal-fired power stations, justifying the move by referring to the number of people who urgently need to be pulled out of poverty; coal still remains the cheapest option for fulfilling this purpose. The Indian government is fond of quoting Gandhi on the present environmental crisis: "Earth [*prithvi*] provides enough to satisfy every man's need but not enough for every man's greed."[41] Yet "greed" and "need" become indistinguishable from each other in arguments in defense of continued use of coal, the worst offender among fossil fuels. India and China want coal; Australia and other countries want to export it. It is still the cheapest variety of fossil fuel. In 2011, "coal represented 30 percent of world energy," and that was "the highest share it [had] had since 1969."[42] Coal use was expected to increase by 50 percent by 2035, bringing enormous export opportunities to companies in South America. "American coal companies," remarked a report in the *New York Times*, "badly want to export coal from the country's most productive mines in the Powder River Basin in Wyoming and Montana" as they saw that in the longer term, thanks to China and India, coal's future seemed "bright—mainly because it is cheaper than its competitors."[43] This vast market for coal would not have come about without China and India justifying the use of coal by referring to the needs of their poor. So it is, as Amitav Ghosh points out in his *The Great Derangement*, the size of the populations of these two nations that gives the climate crisis a distinctly Asian future.[44] The physicist P. W. Anderson famously said in 1972 "more is different."[45] Rapid population growth in already populous societies, as has happened in the world since 1900, changes the relationship between human societies and the biosphere. As many have pointed out, the exponential growth of human population in the twentieth century has itself had much to do with fossil fuels through the use of artificial fertilizers, pesticides, and irrigation pumps.[46]

Population is also a problem in yet another sense. The total size and distribution of humanity matters in how the climate crisis unfolds, particularly with regard to species extinction. There is the widely accepted point that humans have been putting pressure on other species for quite some time now; I do not need to belabor it. Indeed, the war (in spite of traditions of interspecies relatedness) between humans and animals such as rhinoceroses, elephants, monkeys, and big cats may be seen every day in many Indian cities and villages.[47] That we have consumed many varieties of marine life out of existence is also generally accepted. Ocean acidification threatens the lives of many species.[48]

But there is another reason why the history of human evolution and the total number of human beings today matter when we get to the question of species survival as the planet warms. One way that species threatened by global warming will try to survive is by migrating to areas more conducive to their existence. This is how they have survived past changes in the climatic conditions of the planet. But now there are so many of us, and we are so widespread on this planet, that we stand in the way. Curt Stager puts it clearly:

> Even if we take a relatively moderate emissions path into the future and thereby hope to avoid destroying the last polar and alpine refuges, warming on the scale [expected] ... will still nudge many species toward higher latitudes and elevations. In the past, species could simply move ... but this time they'll be trapped within the confines of habitats that are mostly immobilized by our presence.... As Anthropocene warming rises toward its as yet unspecified peak, our long-suffering biotic neighbors face a situation that they have never encountered before in the long, dramatic history of ice ages and interglacials.[49]

They can't move because we humans are standing in their way.

The irony of the point runs deeper. The spread of human groups throughout the world—the most remote Pacific islands were the last to be settled by around 3000 BP[50]—and industrial-age population growth now make it difficult for human climate refugees to move to safer and more inhabitable climes. Other humans will stand in their way. Burton Richter puts the point thus:

> We [humans] were able to adapt to [climate] change in the past ... but there were tens of thousands of years to each swing compared with only hundreds of years for the earth to heat up this time. The slow pace of change gave the relatively small population back then time to move, and that is just what it did during the many temperature swings of the past, including the ice ages. The population now is too big to move en masse, so we had better do our best to limit the damage that we are causing.[51]

The history of population thus belongs to two histories at once: the very short-term history of the industrial way of life—of modern medicine, technology, and fossil fuels (fertilizers, pesticides, irrigation)—that accompanied and enabled the growth in our numbers and life expectancies, and the much, much longer-term evolutionary or deep history of our species, the history through which we have evolved to be the dominant species of the planet, spreading all over it and now threatening the existence of many other life-forms. The poor participate in that shared history of human evolution just as much as the rich do. Add to this Peter Haff's argument about the technosphere that we discussed in the introduction to this book. Minus the network of connections that the technosphere represents, the total human population on earth, he argues, will collapse dramatically. The "technosphere" has become the condition of possibility enabling both the rich and the poor to live on this planet and act as its dominant species.[52]

The per capita emission figures, while useful in making a necessary and corrective polemical point in the political economy of climate change, hide the larger history of the species of which both the rich and the poor partake, albeit differently. Population is clearly a category that joins together the short-term history of iniquitous modernizations and the much longer-term history of the relationship between us *Homo sapiens* and other species.

Are Humans Special? The Moral Rift of the Anthropocene

The climate crisis reveals the sudden coming together—the enjambment, if you will—of the usually separated syntactic orders of recorded and deep histories of the human kind, of species history and the history of the Earth systems, revealing the deep connections through which the planetary processes and the history of biological life interact with each other. From this knowledge it does not follow, however, that humans will stop pursuing, with vigor and vengeance, our all-too-human ambitions and squabbles that unite and divide us at the same time. Will Steffen, Paul Crutzen, and John McNeill have drawn our attention to what they call—after Polyani—the period of "the great acceleration" in human history circa 1945 to 2015, when global figures for population, real GDPs, foreign direct investment, damming of rivers, water use, fertilizer consumption, urban population, paper consumption, transport motor vehicles, telephones, international tourism, and McDonald's restaurants (yes!) all began to increase dramatically in an exponential fashion.[53] This period, they suggest, could be a strong candidate for an answer to the question, When did the Anthropocene begin? The Anthropocene may well stand for a multitude of environmental problems we face today collectively, but it is impossible for me, as a historian of human affairs, not to notice that this period of so-called great acceleration is also the period of great decolonization in countries that had been dominated by European imperial powers and that made a move toward modernization (the damming of rivers, for instance) over the ensuing decades and, with the globalization of the last twenty years, toward a certain degree of democratization of consumption as well. I cannot ignore the fact that "the great acceleration" included the production and consumption of consumer durables—such as the refrigerator and the washing machine—in Western households that were touted as "emancipatory" for women.[54] Nor can I forget the pride with which today the most ordinary and poor Indian citizen possesses his or her own smart phone or its cheap substitute.[55] The lurch into the Anthropocene has also been globally the story of some long-anticipated social justice, at least in the sphere of consumption.

This justice among humans, however, comes at a price. The result of growing human consumption has been a near-complete human appropriation of the biosphere. Jan Zalasiewicz cites some sobering statistics from the researches of Vaclav Smil:

> Smil has taken our measure from the most objective criterion of all: collective weight. Considered simply as body mass . . . we now bulk up to about a third of terrestrial vertebrate body mass on Earth. Most of the other two-thirds, by the same measure, comprise what we keep to eat: cows, pigs, sheep and such. Something under 5% and perhaps as little as 3%, is now made of the genuinely wild animals—the cheetahs, elephants, antelopes and the like. . . . Earlier in the Quaternary [the last two million years], . . . humans were just one of some 350 large . . . vertebrate species.

"Given the precipitate drop in the numbers of wild vertebrates, one might imagine that vertebrate biomass as a whole has gone down," writes Zalasiewicz. "Well, no," he continues: "Humans have become very good at, firstly, increasing the rate of vegetable growth, by conjuring nitrogen from the air and phosphorus from the ground, and then directing that extra growth towards its brief stopover in our captive beasts, and thence, to us. . . . The total vertebrate biomass has increased by something approaching an order of magnitude above 'natural' levels (staggering, isn't it . . .)."[56] Smil concludes his massively researched book *Harvesting the Biosphere* with these cautionary words: "If billions of poor people in low-income countries were to claim even half the current per capita harvests prevailing in affluent economies, too little of the Earth's primary production would be left in its more or less natural state, and very little would remain for mammalian species other than ours."[57]

This raises a question that bears striking similarity to the question that Europeans often asked themselves when they forcibly or otherwise took over other peoples' lands: by what right or on what grounds do we arrogate to ourselves the almost exclusive claims to appropriate for human needs the biosphere of the planet? John Broome confronts this question in his book on "ethics in a warming world." In a section titled "What Is Ultimately Good?," Broome acknowledges that climate change raises this question: "in particular the question if nature—species, ecosystems, wildernesses, landscapes—has value in itself." That question he decides is "too big" for his book and yet still proceeds to offer these thoughts on the value of nature: "Nature is undoubtedly valuable because it is good for people. It provides material goods and services. The river brings us our clean water and takes away our dirty water. Wild plants provide many of our medicines.... Nature also brings emotional good to people. But the significant question raised by climate change is whether nature has value in itself.... This question is too big for this book. I shall concentrate on the good of the people."[58]

But is "the good of the people" an unquestionable good? Are we special? Archer also begins his book *The Long Thaw* addressing this very question. Science, Archer thinks, is humbling for humans, for it does not hold up the case for human specialness. It rather tells us we are not "biologically 'special'"—"we are descended from monkeys, and they from even humbler origins." Geological evidence, he further writes, "tells us that the world is much older than we are, and there's no evidence that it was created especially for us.... This is all very humbling."[59] But the tricky question of the assumed specialness of humans takes us into a past much longer than that of capital and into territories that we never had to cross in thinking about the inequalities and injustices of the rule of capital.

The idea that humans are special has, of course, a long history. We should perhaps speak of anthropocentrisms in the plural here. There is, for instance, a long line of thinking—from religions that came long after humans established the first urban centers of civilization and created the idea of a transcendental God through to the modern social sciences—that has humans opposed to the natural part of the world. These later religions are in strong contrast, it seems, with the much more ancient religions of hunting-gathering peoples (I think here of the Australian Aboriginals and their stories) that often saw humans as part of animal life (as though we were part of *Animal Planet* and not simply watching it from outside the idiot box). Humans were not necessarily special in these ancient religions. Recall Émile Durkheim's position on totemism. In determining "the place of man" in the scheme of totemistic beliefs, Durkheim was clear that totemism pointed to a doubly conceived human or what he called the "double nature" of man: "Two beings co-exist within him: a man and an animal." And again: "we must be careful not to consider totemism a sort of animal worship.... Their [men and their totems] relations are rather those of two things who are on the same level and of equal value."⁶⁰

The very idea of a transcendental God puts humans in a special relationship with the Creator and his creation, the world. This point needs a separate and longer discussion, but for a completely random and arbitrary—arbitrary, for I could have chosen examples from other religious traditions, including Hinduism—example of this for now, consider the following remarks from Fazlur Rahman. By way of explaining the term *qadar*—meaning both "power and measuring out"—that the Qur'an uses in close association with another word, *amr*, meaning "command," to express the nature of God, Rahman remarks thus on God's relationship to man as mediated through nature:

> The all-powerful, purposeful, and merciful God ... "measures out" everything, bestowing upon everything the right range of its potentialities, its laws of behavior, in sum, its character. This measuring on the one hand ensures the orderliness of nature and on the other expresses the most fundamental, unbridgeable difference between the nature of God and the nature of man: the Creator's measuring implies an infinitude wherein no measured creature ... may literally share.

This is why "nature does not and cannot disobey God's commands [*amr*] and cannot violate natural laws."⁶¹ While this enjoins very clearly that man must not play God, it does not mean, as Rahman clarifies, that "man cannot discover those laws and apply them for the good of man."⁶² God is kind because he has stocked the world with provisions for us!⁶³ Environmentalists, similarly, have long cited a verse in Genesis in which "the Lord says '[Let men] have dominion ... over all the earth, and over every creeping thing that creeps on earth.' He enjoins man to 'be fruitful and multiply and fill the earth and subdue it.'"⁶⁴

The literature on climate change thus reconfigures an older debate on anthropocentrism and so-called nonanthropocentrism that has long exercised philosophers and scholars interested in environmental ethics: do we value the nonhuman for its own sake or because it is good for us?⁶⁵ Nonanthropocentrism, however, may indeed be a chimera, for as Feng Han points out in a different context, "human values will always be from a human (or anthropocentric) point of view."⁶⁶ While ecologically minded philosophers in the 1980s made a distinction between "weak" and "strong" versions of anthropocentrism, they supported the weaker versions. Strong anthropocentrism had to do with unreflexive and instinctive use or exploitation of nature for purely human preferences; weak anthropocentrism was seen as a position arrived at through rational reflection on why the nonhuman was important for human flourishing.⁶⁷

Lovelock's work on climate change, however, produces a radically different position, on the other side of the rift as it were. He packs it into a pithy proposition that works almost as the motto of his book *The Vanishing Face of Gaia*: "to consider the health of the Earth without the constraint that the welfare of humankind comes first."⁶⁸ He emphasizes, "I see the health of the Earth as primary, for we are utterly dependent upon a healthy planet for survival."⁶⁹ What does it mean for humans, given their inescapable anthropocentrism, to consider "the Earth as primary" or to contemplate the implications of Archer's statement that the world was not "created especially for us"? I will consider this question in the following and concluding section of this essay and then follow up on it in subsequent chapters.

Climate and Capital, the Global, and the Planetary

All of the rifts I have discussed here turn on the difference between human time and the deeper and longer temporal rhythms of the geobiological processes that contribute to the making of Earth system history. Whether we stay with this difference or try to fold it back into the temporality of human institutions and their history is the question that constitutes the nub of the debate in which this book is situated. In *Living in the End Times*, Slavoj Žižek critiqued the arguments I presented as I started working on this project. Some of his comments concern points about the "true" nature of Hegelian dialectic, which I will not discuss here. But he also made a point about the relationship between anthropogenic climate change and "the capitalist mode of production" that allows me to get into my final stride here. Responding to my points that there were "natural parameters" to our existence as a species that were relatively independent of our choices between capitalism and socialism and that we therefore needed to think deep history of the species and the much shorter history of capital together, Žižek remarked,

> Of course, the natural parameters of our environment are "independent of capitalism or socialism"—they harbor a potential threat to all of us, independently of economic development, political system, etc. However, the fact that their stability has been threatened by the dynamic of global capitalism nonetheless has a stronger implication than the one allowed by Chakrabarty: in a way, we have to admit that the Whole is contained by its Part, that the fate of the Whole (life on earth) hinges on what goes on in what was formerly one of its parts (the socio-economic mode of production of one of the species on earth).

Given this premise, his conclusion followed:

> [We also] have to accept the paradox that . . . the key struggle is the particular one: one can solve the universal problem (of the survival of human species) only by first resolving the particular deadlock of the capitalist mode of production. . . . The key to the ecological crisis does not reside in ecology as such.[70]

Žižek's proposition with regard to the role of the capitalist mode of production in the drama of climate change goes well beyond what I have proposed in this chapter. That capitalist or industrial civilization, dependent on large-scale availability of cheap fossil-fuel energy, is a proximate or efficient cause of the climate crisis is not in doubt. I am in agreement with most scholars on that point. But Žižek puts only capitalism in the driver's seat; it is the "part" that now determines "the whole."

Ursula Heise has pointed out sharply why Žižek's dialectics are simply unhelpful in dealing with the crisis of global warming. Planetary warming, she writes, "will not stop tomorrow: even if a collective will to develop an alternative economic regime were to emerge in some of the planet's dominant nations, the transition to such a regime would almost certainly take decades (more likely, a century or more) — too late to affect the current climate crisis decisively. Žižek's assumption that overcoming capitalism is a prerequisite for addressing the climate crisis, in practice, simply denies the possibility of coming to terms with it."[71]

There is, besides, a larger problem with Žižek's understanding itself: to say that the history and logic of a particular human institution has gotten caught up in the much larger processes of the Earth systems and evolutionary history (stressing the lives of several species including ourselves) is not to say that human history is the driver of these large-scale processes. These latter processes continue over scales of space and time that are much larger than those of capitalism — hence the rifts we have discussed. As Stager and Archer point out, however much "excess" CO_2 we put out today, the long-term processes of the Earth system, its million-year carbon cycle, for instance, will most likely "clean it up" one day, humans or no humans.[72] Which is why it seems logically more consistent to see these long-term Earth system processes as co-actors in the drama of global warming. This is also suggested by the fact that, unlike the problems of wealth accumulation or income inequalities or the questions posed by globalization, the problem of anthropogenic climate change could not have been predicted from within the usual frameworks deployed to study the logics of capital. The methods of political-economic investigation and analyses do not usually entail digging up 800,000-year-old ice-core samples or making satellite observations of changes in the mean temperature of the planet's surface. Climate change is a problem defined and constructed by climate scientists whose research methods, analytical strategies, and skill sets are different from those possessed by students of political economy.

Once we grant processes belonging to the deeper histories of Earth and life the role of coactors in the current crisis, playing themselves out on both human and unhuman scales, the prescience of a sentence Gayatri Chakravorty Spivak wrote a while ago comes into view: "The planet is in the species of alterity, belonging to another system; and yet we inhabit it."[73] Spivak was on to something. Her formulation takes a step toward pondering the human implications of the planetary studies that inform and underpin the science of climate change.

This science helps us develop an emergent conception of the planetary that is related to but different from the existing conceptions of the global. For even though the current phase of warming of the earth's atmosphere is indeed anthropogenic, it is only contingently so; humans have no intrinsic role to play in the science of planetary warming as such. The science is not even specific to this planet; it is part of what is called planetary science. It does not belong to an earthbound imagination. A textbook used in many geophysics departments to teach planetary warming is simply called *Principles of Planetary Climate*.[74] Our current warming is an instance of planetary warming that has happened both on this planet and on other planets, humans or no humans, and with different consequences. It just so happens that the current warming of the earth is of human doing.

The scientific problem of climate change thus emerges from what may be called comparative planetary studies and entails a degree of interplanetary research and thinking. The imagination at work here is not human centered. It speaks to a growing divergence in our consciousness between the global—a singularly human story—and the planetary, a perspective to which humans are incidental.[75] The climate crisis is about waking up to the rude shock of the planet's otherness. The planet, to speak with Spivak again, "is in the species of alterity, belonging to another system." And "yet," as she puts it, "we inhabit it." If there is to be a comprehensive politics of climate change, it has to begin from this perspective. The realization that humans—all humans, rich or poor—come late in the planet's life and dwell more in the position of passing guests than possessive hosts has to be an integral part of the perspective from which we pursue our all-too-human but legitimate quest for justice on issues to do with the iniquitous impact of anthropogenic climate change.

* 3 *
The Planet

A HUMANIST CATEGORY

Earth System Science (ESS), the science that among other things explains planetary warming and cooling, gives humans a very long, multilayered, and heterotemporal past by placing them at the conjuncture of three (and now variously interdependent) histories whose events are defined by very different timescales: the history of the planet, the history of life on the planet, and the history of the globe made by the logics of empires, capital, and technology. One can therefore read Earth system scientists as historians writing within an emergent regime of historicity. We could call this a planetary or anthropocenic regime of historicity to distinguish it from the global regime of historicity that has enabled many humanist and social-science historians to deal with the theme of climate change and the idea of the Anthropocene. In the latter regime, however, historians try to relate the Anthropocene to histories of modern empires and colonies, the expansion of Europe and the development of navigation and other communication technologies, modernity and capitalist globalization, and the global and connected histories of science and technology.[1]

It is my contention that when we read together—as we must—histories produced on these two registers, the category *planet* emerges as a category of humanist thought, a category of existential and, therefore, philosophical concern to humans. Martin Heidegger pronounced the word *planet* as being of no interest to philosophers when he introduced *earth* as a philosophical category in 1936, distinguishing it carefully from the word *planet*. "What this word [earth] says," he wrote, "is not to be associated with the idea of a mass of matter deposited somewhere, or with the merely astronomical idea of a planet."[2] His lecture on "The Origin of the Work of Art," delivered first in Frankfurt that year, explained "earth" as that which made life possible. It was the ground for humans' attempt to dwell: "Upon the earth and in it, historical man grounds his dwelling in the world."[3] Or, as he put it in another essay, "Earth is the serving bearer, blossoming and fruiting, spreading out in rock and water, rising up into plant and animal."[4] When mortals dwelled on earth, they "saved" it. "Saving," Heidegger explained, "does not only snatch something from a danger.... To save the earth is more than to exploit it or even wear it out. Saving the earth does not master the earth and does not subjugate it, which is merely one step from spoliation."[5] Human worlds and the earth are in a relationship of strife — that is, it is never simply a relationship of harmony and can give rise to anxiety, for instance, as part of dwelling — and are yet mutually bonded. "World and earth are essentially different from one another," writes Heidegger, "and yet are never separated. The world grounds itself on the earth, and earth juts through the world.... The opposition of world and earth is a striving."[6]

Heidegger's turn toward philosophizing the earth produced a minor intellectual tumult among his followers. In "The Truth of the Work of Art" (1960), Hans-Georg Gadamer remembered what "a new and startling thing" it was to have the category "earth" thus introduced as a foil to Heidegger's concept of the "world."[7] Fourteen years later, writing on the occasion of Heidegger's eighty-fifth birthday, Gadamer returned to this subject and mentioned how "quite unusual" it was "to hear talk of the earth and the heavens, and of a struggle between the two — as if these were concepts of thought that one could deal with in the same way that the metaphysical tradition had dealt with the concepts of matter and form."[8]

The earth/world distinction and the earth/planet distinction cut in different ways for Heidegger's readers today. If his earth/world distinction helped him formulate his ideas on human dwelling, his earth/planet distinction, by contrast, roughly maps onto the division that some Earth system scientists make between the zone of the planet that is critical to the maintenance of life — the critical zone, as it is called — and the rocky, hot, and molten interior of the planet. The "critical zone" is "Earth's near surface layer from the tops of the trees down to the deepest groundwater, where most human interactions with the Earth's surface take place and [which is] the locus of most geomorphological activity."[9] Using Heidegger's language, we can say that the harder we work the earth in our increasing quest for profit and power, the more we encounter the planet. *Planet* emerged from the project of globalization, from "destruction" and the futile project of human mastery (what Heidegger would call "impotence of will").[10] Yet it is neither the globe nor the world and definitely not the earth. It belongs to a domain where this planet reveals itself as an object of astronomical and geological studies and as a very special case containing the history of life — all of these dimensions vastly out-scaling human realities of space and time.

A profound difference separates the planet from the three categories we have thought with so far in thinking world or global history: world, earth, and globe (sometimes treated as synonymous with the planet). These are all categories that, in various ways, reference the human. They have this orientation in common. We see the globe as created by human institutions and technology. Humans and earth, as Heidegger saw it, stand in a face-to-face relationship.[11] In Heidegger's thought, the earth had to wait as it were for the coming of language, for it was only when a creature evolved that was capable of using language that the question of being — the meaning of having to be — could be vouchsafed to it.[12] But the planet is different. We cannot place it in a communicative relationship with humans. It does not as such address itself to humans, unlike, say, the Heideggerian "earth" — or maybe even James Lovelock's or Bruno Latour's Gaia — that does.[13] To encounter the planet in thought is to encounter something that is the condition of human existence and yet remains profoundly indifferent to that existence.

Humans have empirically encountered the planet — deep earth — always in their history — as earthquakes, volcanic eruptions, and tsunamis — without necessarily encountering it as a category in humanist thought. They have — as shown by Voltaire's debate with the dead Gottfried Wilhelm Leibniz after the 1755 earthquake in Lisbon or by Mahatma Gandhi's debate with Rabindranath Tagore after the 1934 earthquake in Bihar — dealt with the planet without having to call it by that name.[14] The planet was folded into human debates about morality, theodicy, and more recently into the idea of natural disaster.[15] But as evidence gathers that the nature/human distinction is, ultimately, unsustainable and that human activities worldwide may even contribute to the increasing frequency of earthquakes, tsunamis, and other "natural" disasters, the planet *as such* has emerged as a site of existential concern for those who write its histories in what I have called the planetary or anthropocenic regime of historicity. These are none other than Earth system scientists themselves. Their accounts show the Earth system to be in danger of being gravely disturbed — these histories have bared the planet as an entity to reckon with in debating human futures. *Planet* is not a lazy word in these narratives. It is a dynamic ensemble of relationships — much as G. W. F. Hegel's *state* or Karl Marx's *capital* were — an ensemble that constitutes the Earth system. It is at such moments of concern expressed by scientists over the state of the Earth system that the planet (i.e., Earth system) emerges as a category of humanist thought. Heidegger's stance against science and his assumption that the nature of human dwelling can be imagined without thinking of the "astronomical" object, our planet, are positions we cannot support in the time of the Anthropocene.

The nature of this new category *planet* is best explored, it seems to me, by distinguishing it from the idea of the globe with which it has often been identified in the past. I begin by elaborating on this distinction between the globe and the planet. The category *earth*—relevant to this exercise but not directly addressed here—contains a further distinction between the land and the sea that, as we will see, remained central to Carl Schmitt's thoughts on human dwelling that I want to draw on to frame my overall argument.[16] I am, of course, not the first person to take a planetary turn. My thoughts on the globe/planet distinction began in the previous chapter in an encounter with Gayatri Chakravorty Spivak's invocation of planetarity, though, as readers will see, I have now pushed them in a particular direction.[17]

The Global and the Planetary: The Globe of Globalization[18]

The word *globe* as it has appeared in the literature on globalization is not the same as the word *globe* in the expression *global warming*.[19] The story of globalization has humans at its center and narrates how humans historically connected into a human sense of the globe. Fields like world history and global history, for all their differences, have contributed to our understanding of this process. Take two texts, separated by more than three hundred years—Thomas Hobbes's *Leviathan* (1651) and Hannah Arendt's *The Human Condition* (1958)—one inaugurating modern political thought, the other renewing political philosophy at a time when space travel had just begun. Notice how much their sense of what the earth was for humans ("knowledge of the face of the earth") was conditioned, even across centuries, by the history of European expansion, trade, the mapping and navigation of the seas (and eventually the air), along with the development of instruments of navigation and mobility—in other words, processes and institutions that created the modern sense of the globe.[20] It is as if Hobbes's historical references, in one of his most remembered passages, describing how the condition of humans changed with the rise of the state—"In such condition [the state being absent], there is no place for Industry; because the fruit thereof is uncertain: and consequently no Culture of the Earth [earth here understood as *land to be cultivated*]; no Navigation, nor use of the commodities that may be imported by Sea; . . . no Instruments of moving, and removing such things as require much force; no Knowledge of the face of the Earth; no account of Time"—repeated themselves verbatim as Arendt positioned herself in the late 1950s, observing the same historical process that Hobbes had seen in an earlier phase of its development.[21] "As a matter of fact," she wrote,

> The discovery of the earth, the mapping of her lands and the chartering of her waters [once again the land/sea distinction], took many centuries and has only now begun to come to an end. Only now has man taken full possession of his mortal dwelling place and gathered the infinite horizons . . . into a globe whose majestic outlines and detailed surface he knows as he knows the lines in the palm of his hand. Precisely when the immensity of available space on earth was discovered, the famous shrinkage of the globe began, until eventually in our world . . . each man is as much an inhabitant of his earth as he is an inhabitant of his country. Men now live in an earth-wide continuous whole. . . . Nothing, to be sure, could have been more alien to the purpose of the explorers and circumnavigators of the early modern age than the closing-in process; they went to enlarge the earth, not shrink her into a ball. . . . Only the wisdom of hindsight sees the obvious, that nothing can remain immense if it can be measured.[22]

These quotations from two fundamental thinkers in the European tradition show how central the story of European expansion is to their narratives of the making of the globe.

Schmitt's *The Nomos of the Earth*, though relatively old, is still percipient enough to give us a handle over the history of this particular version of the globe. Schmitt tells a story of how the idea of law got dislodged from its association with earth, understood as land and dwelling, when the seas opened up to an expanding and imperial Europe. *Nomos* (law) was originally land bound and was about *appropriation* of land, a process that Schmitt argued was profoundly connected to a fundamental human orientation to land and territory (as seen most clearly in the case of Australian Aboriginals, say), and thus to strife and war between humans over appropriation of land.[23] The sea was just an extensive surface that did not allow for boundaries; all human ideas about *nomos* were firmly grounded in the occupation of particular patches of land and thus to the practice of erecting boundaries. Schmitt even cites a Biblical passage showing a human imagination of an ideal planet that had no sea.[24] It was only when appropriation of land was secured—by "migrations, colonizations, and conquests"—that humans could engage in the processes required for social formation: "distribution," by which Schmitt meant the setting up of an order, and "production," which referred to the organization of the economic life of a society.[25]

Thus, in Schmitt's schema, the chain of logic went like this: appropriation→distribution→production. The sense of being at home in a particular place could come about only after the process of appropriation had been completed. Appropriation was therefore related to the idea of dwelling. Yet, as Schmitt writes, "the distribution remains stronger in memory than does the appropriation, even though the latter was the precondition of the former."[26] However, Schmitt adds, this land-bound sense of "the first *nomos* of the world was destroyed about 500 years ago when the great oceans were opened up."[27]

Nomos gradually ceased to be something land-based and thus orienting for humans. It lost its connection to dwelling. There came about a separation, at the intellectual level of jurisprudential thought, between the ought and the is, between *nomos* and *physis* (this separation being the precondition for, among other things, international law). The coming of air travel and eventually the space age would only expand this separation of *nomos* and *physis* and leave humans with two options in the future: either feeling "homeless" (as the globe is home for nobody) or working toward a unity in which all humans come to regard the globe as their home.

Most histories of globalization assume—to stay with Schmitt's schema—that the struggle between humans for appropriation of land, sea, or space is now over. Humans are now spread all over the globe; there is nowhere else to go; we control the skies and the waters. We are in a postimperial age, on this account, so our struggle is in the sphere of what Schmitt called "distribution"—that is, about establishing a just order so that the idea of *nomos* continues to remain unrelated to *physis*. Many climate-justice arguments, for instance, relate to a just distribution of an abstract and global carbon space. The particular niceties of Schmitt's argument are not my concern here—except that a climate-ravaged world with migrants and refugees can reopen arguments about appropriation. The point relevant here is that in Schmitt's and others' histories of globe making, the words *planet* and *globe* remain synonymous, as Schmitt's own usage reveals:

> The first attempts in international law to divide the earth as a whole according to the new global concept of geography began immediately after 1492. These were also the first adaptations to the new, planetary image of the world.[28]

> The compound term "global linear thinking" . . . is also better than "planetary" or similar designations, which refer to the whole earth, but fail to capture its characteristic type of division.[29]

> The English island [at the time of the Treaty of Utrecht in 1713] remained a part of or rather the center of this European planetary order.³⁰

> I speak of a new *nomos* of the earth. That means that I consider the earth, the planet on which we live, as a whole, as a globe, and seek to understand its global division and order.³¹

This mode of equating the planet with the globe remained with Schmitt even in his later texts, such as *Land and Sea*: "As [the nineteenth-century German geographer Ernst] Kapp remarked, the compass lent the ship a spiritual dimension which enabled man to develop a strong attachment to his ship, a sort of affinity or kinship. From then on, the remotest oceanic lands could come into contact with each other, and the planet opened itself to man."³² Here "planet" was simply another word for *globe*; it referred to the planet we live on, the earth taken "as a whole."

The same is true, incidentally, of Heidegger's use—when he actually did use them—of the words *planet* or *planetary*. The expression "planetary imperialism" turns up towards the very end of Heidegger's "The Age of the World Picture," which has influenced much recent thinking on images of the earth taken from space.³³ He writes, "In the planetary imperialism of technologically organized man, the subjectivism of man attains its acme, from which point it will descend to the level of organized uniformity and there firmly establish itself. This uniformity becomes the surest instrument of total, i.e., technological, rule over the earth."³⁴ "Planetary" refers here to the earth as a single planet *taken by itself*, not studied in comparison to other planets. This becomes obvious also from the way Heidegger, in another essay, assigns the "planet" to an "advancing world history."³⁵ Since both imperialism and world history are categories of human history, the word *planet* in Heidegger's usage refers to nothing other than the globe. In fact, it is the connection he makes between "man's" "planetary imperialism," "his" technological rule, and the rising to a peak of "man's" subjectivism that allows Heidegger to develop a critique of this "planetary imperialism" in a way that generates in turn a powerful critique of a certain dominant "anthropology" (Heidegger's word):

> When the world becomes picture, what is, in its entirety, is juxtaposed as that for which man is prepared and which, correspondingly, he therefore intends to bring before himself and have before himself, and consequently intends in a decisive sense to set in place before himself. ... The Being of whatever is, is sought and found in the representedness of the latter.³⁶

The globe of globalization embodies this anthropocentric and anthropological practice of representation.

The Global and the Planetary: The Globe of Global Warming

Anthropogenic global warming is no doubt connected to the story of globalization. One could even argue that a certain period in the history of globalization now known as "the great acceleration" (1950 onward) overwhelmingly contributed to the forging of this connection, so much so that some scholars have pinned the beginning of the Anthropocene down to this period itself.[37] But the science of global warming takes us away from an earth- and human-bound imagination. For this reason it also effects a profound unsettling of the narrative of globalization. Earth System Science (ESS) is a mode of looking at this planet that, in contrast to the globe of globalization, *necessarily has other planets in view* in order to create models of how this planet works (and the principles of representation involved are different from those involved in invoking the globe). Contrary to what we might imagine, the science of global warming is not even specific to this planet — it is part of what is called planetary science.[38] Indeed, our current warming is simply an instance of what is called planetary warming. Such warming has happened both on this planet and on other planets with widely different consequences. It just so happens that the current warming of the earth is primarily a result of human actions.

It is not at all an accident that two of the foundational scientists associated with this science — James Lovelock and James Hansen — began their careers, respectively, by being associated with the study of Mars and Venus. Hansen was initially a student of planetary warming on Venus and only later transferred his interests to earth out of concern and curiosity. Hansen writes, "In 1978, I was still studying Venus." He shifted to studying the earth because, he says,

> The atmosphere of our home planet was changing before our eyes, and it was changing more and more rapidly.... The most important change was the level of carbon dioxide, which was being added to the air by the burning of fossil fuels. We knew that carbon dioxide determined the climate on Mars and Venus. I decided it would be more useful and interesting to try to help understand how the climate of our own planet would change, rather than study the veil of clouds shrouding Venus.

He shifted the site of his research to this planet thinking, he writes with an obvious touch of irony, that it would be a "temporary obsession."[39]

ESS was a product of the Cold War and the military and civil competition that it spawned in space. This history has been recounted by

Joshua Howe, Spencer Weart, and more recently by Ian Angus and Clive Hamilton, and it need not be repeated here in detail.[40] While some of the basic ideas related to ESS go back to the nineteenth and early twentieth centuries, NASA first set up its ESS committee in 1983 when it realized that the planet needed to be studied as a whole by different kinds of scientists.[41] It is a deeply interdisciplinary science, synthesizing "elements of geology, biology, chemistry, physics, and mathematics."[42] The International Geosphere-Biosphere Programme, launched in 1987, defined *Earth system* as follows:

> The term "Earth system" refers to Earth's interacting physical, chemical, and biological processes. The system consists of the land, oceans, atmosphere and poles. It includes the planet's natural cycles—the carbon, water, nitrogen, phosphorus, sulphur and other cycles—and deep Earth processes. Life too is an integral part of the Earth system. Life affects the carbon, nitrogen, water, oxygen and many other cycles and processes. The Earth system now includes human society. Our social and economic systems are now embedded within the Earth system. In many cases, the human systems are now the main drivers of change in the Earth system.[43]

Will Steffen, an Earth system scientist, thus described the intellectual ambit of this emergent science:

> Crucial to the emergence of this perspective has been the dawning awareness of two fundamental aspects of the status of the planet. The first is that the Earth itself is a single system, within which the biosphere is an active, essential component.... Second, human activities are now so pervasive and profound ... that they affect the Earth at a global scale in complex, interactive, and accelerating ways ... that threaten the very processes and components, both biotic and abiotic, upon which humans depend.[44]

System is used in the singular in ESS to underscores the systemic nature of the planetary processes under study.

Bruno Latour and Tim Lenton have recently raised the question of whether the so-called Earth system is indeed one system or if we should even think about it as constituting "a whole."[45] To my nonspecialist ears, their question certainly sounds legitimate. I do not know whether multiple, different, and yet interacting flows and feedback loops in earth processes do indeed constitute a *single* system. But it has to be noted that this position is somewhat in tension with Lenton and Andrew Watson's statement that "the many processes that interact together to set the living conditions at the surface of the planet" constitute "a very coherent system."[46] There are clearly some very widely shared working agreements among scientists in this area as well as some major differences indicating, perhaps, how young this interdisciplinary science still is. In his introductory book on ESS, Lenton, for example, writes about the "fuzzy lower boundary to the Earth system":

> The temptation is to include the whole of the interior of the planet in the Earth system—and this is exactly what NASA's 1986 report did when considering the longest timescales. . . . However, for many Earth system scientists, the planet Earth is really composed of two systems—the surface Earth system that supports life, and the great bulk of the inner Earth underneath.

Lenton focuses deliberately on "the thin layer of a system at the surface of the Earth—and its remarkable properties," the critical zone that I mentioned above.[47] Lee R. Kump, James F. Kasting, and Robert G. Crane's *The Earth System*, on the other hand, deals with what the authors regard as "four parts" of the Earth system: the atmosphere, the hydrosphere, the biota, and the solid earth. What their text helps to clarify is that this new science is as much about taking a systems approach to the study of how the earth "works" as it is about observing how "the processes active on Earth's surface are *functioning together* to regulate climate, the circulation of the ocean and atmosphere, and the recycling of the elements [such as carbon, nitrogen, oxygen, and more]" with the biota—life—playing "an important role in all these processes."[48]

The deeper parts of the planet affect the biosphere for sure (as plate tectonics does, for example, or volcanic eruptions do) and are fundamentally important in supplying geochemically fresh landscapes; the question is whether they constitute parts of the Earth system.[49] However this is resolved, there is no denying that planetary processes operating on different scales and involving the actions of both the living and the nonliving are often interlocked in complicated, complex, and precarious ways, and it is the fact of their being interlocking and interactive in character that is highlighted by the use of the term *Earth system*. For Erle C. Ellis, observations and computer modeling of the Earth system clearly documented in the 1990s that "human activities were changing in tandem with changes in Earth's atmosphere, lithosphere, hydrosphere, biosphere, and climate," leading scientists and others experts associated with the International Geosphere-Biosphere Programme to announce in one voice in 2001—this is known as the Amsterdam Declaration on Global Change—that "the Earth system behaves as a single, regulating system comprised of physical, chemical, biological and human components."[50] It is somewhat odd that this declaration should have separated the "human component" from the physical, chemical, and biological ones, but clearly a political point was made by such a separation.

The immediate roots of this interdisciplinary science, as I have mentioned before, go back to the Cold War years of the 1960s when Lovelock, working for Carl Sagan's unit in NASA, developed his now-famous ideas regarding Gaia proposing that life on Earth created the conditions for its continued maintenance, as though the earth behaved as a single superorganism that he christened, on advice from William Golding, Gaia.[51] The concept was further developed in the 1970s by Lynn Margulis. Lovelock's early homeostatic view of the planet did not survive scientific skepticism, but his fundamental question as to what made the earth so continuously habitable for life, something the two neighboring planets Mars and Venus were not, survived into ESS as the so-called habitability problem that today is central, for instance, to disciplines like astrobiology or to the search for earthlike exoplanets in the universe.

The important point for our discussion is that the chief protagonist of the story that ESS tells is not humans or human life but complex, multicellular life in general. In contrast to the story of capitalist globalization, this outlook lays out a perspective on humans and other forms of life without humans being at the center of the story. We simply come too late in the story to be its protagonist. This science, of course, is produced by humans and therefore practices a human version of non-anthropocentrism, an attempt by humans to understand their own story by standing outside, as it were, of the story of humans (as the historical sciences of geology and evolutionary biology routinely do). Besides, as Lovelock himself pointed out, ESS entails a view of the planet that is essentially taken from the outside. Lovelock wrote, "To my mind, the outstanding spin-off from space research is not new technology. The real bonus has been that for the first time in human history we have had a chance to look at the earth from space, and the information gained from *seeing from the outside* our azure-green planet in all its global beauty has given rise to whole new set of questions and answers."[52]

Lovelock was right to say that space travel afforded humans a chance to view the planet from outside, but we should note that while this was indeed the first time some humans actually saw their planet as a whole, humans have imagined the planet from the outside for a long time, at least in European history. Ayesha Ramachandran's *The World-makers* presents a fascinating study of this aspect of European imagination in the sixteenth century. Gerhard Mercator's *Atlas*, writes Ramachandran, "define[d] an intellectual watershed by seeking to envision the totality of the world." His 1569 navigational projection still provides the "the basis" for the "Web Mercator platform used by Google Maps and in ArcGIS systems today."[53] Influential in this tradition was also the later Christianized but originally Stoic conception of *kataskopos*—the imaginary "360-degree 'view from above' . . . through which man could transform himself from being a prisoner within the world to becoming a spectator from without"—that was disseminated in renaissance Europe by the popular, fifth-century commentary by Macrobius on Cicero's *Somnium Scipionis*, a part of Cicero's *De re publica* (54–51 BC).[54] It described the Roman general Scipio Aemilianus dreaming of himself looking down on the earth from the starry sphere above.

These were, however, attempts to imagine the earth as it might have appeared to the naked human eye placed somewhere in the sky. One could argue that images of the earth beamed back from space by modern space travelers represent a point of culmination in this history.[55] What distinguishes the "new set of questions" that Lovelock speaks of is that they did not arise from a simple naked-eye view, imagined or real, of the planet from space. The question as to why "since plants and especially forests became established on the land surface, around [more than] 370 million years ago, oxygen has remained between about 17% and 30% of the atmosphere" could not have been raised or answered without asking questions of physics, chemistry, geology, and biology and without comparing this planet with planets like Mars and Venus.[56] To quote Lovelock again, "Thinking about life on Mars gave some of us a fresh standpoint from which to consider life on Earth and led us to formulate a new, or perhaps revive a very ancient, concept of the relationship between the Earth and its biosphere."[57] The planetary is a necessarily comparatist enterprise.

In other words, the Earth system of ESS is produced not simply by a physical view of the planet from outside but by reconstituting it into an abstract figure in the imagination with the help of the sciences—including information obtained from satellites positioned in space as well as from ancient ice-core samples—*while keeping other planets always in view even if only implicitly*. ESS produces a reconstituted planet, the Earth system, an entity no one ever encounters physically but that is, in Timothy Morton's terms, an interconnected series of "hyperobjects"— such as a planetary climate system—(re)created by the use of big data.[58] Delf Rothe has aptly remarked that the Anthropocene is both withdrawn from and inaccessible to earthlings like humans: It is, writes Rothe, "equally totalising and withdrawn: [it] is a new planetary real— a state-shift of the entire Earth System that cannot be known or sensed directly."[59]

There remains, therefore, an interesting tension between ESS and the idea of Gaia. Lovelock was never happy with the name ESS, which he found "anodyne" (see n. 51 above), while Lenton and Watson begin their book with the comment, "'Gaia' and the 'Earth system' are for us, close to being synonymous. . . . [But] 'Earth system science' . . . is . . . less personalized and polarized."[60] Earth System Science is a positive science made up of observed and simulated data and their analyses, but a certain moment of scientific-poetic intuition, such as the moment when the idea later named Gaia flashed through Lovelock's mind, always haunts it.

The Global and the Planetary Diverge

Arendt completed the *The Human Condition* in the shadow of the first artificial satellite, the Soviet Sputnik, adventuring into space. She thought that space technology announced what she referred to as the "earth alienation" of humans, indicating the capacity of the human species to ensure its survival, on other planets if need be, at the great cost of losing their profound sense of being earthbound.[61] A line of famous German thinkers—Spengler, Heidegger, Jaspers, Gadamer, Arendt, and Schmitt among them—watched with foreboding the fast advance of global technology and feared the final "uprooting" of humans, a collapse of the ever-present human project of dwelling by worlding the earth.[62] What we see in the history of ESS, however, is not an end to the project of capitalist globalization but the arrival of a point in history where the global *discloses* to humans the domain of the planetary. We need to keep in mind the poetic nature of Lovelock's vision that constituted the inaugural moment of ESS. True, there had been antecedents of the Gaia theory, but none came with the epiphany of Lovelock's thought about Gaia. Lovelock writes, "The idea of the Earth as a kind of living organism . . . arose in a most respectable scientific environment. . . . It came because my work there led me to look at the Earth's atmosphere from the top down, from space. . . . The air is a mixture that somehow always keeps constant in composition. My flash of enlightenment that afternoon was the thought that to keep [air's composition] constant something must be regulating it and that somehow the life at the surface was involved."[63]

The consciousness that ESS ushers us into simply could not have arisen without the development of technology that "rifled" not only "the bowels of their mother Earth"—as John Milton described early mines—but also the seemingly empty vault of the heavens and all that lies beyond.[64] Consider this: it was the very technology of space exploration that came out of the Cold War and the growing weaponization of atmosphere and space that eventually brought the Gaia moment into our awareness. Or think of our capacity to explore deep earth: climate scientists would not have been able to bore into ice of eight hundred thousand years ago if the US defense establishment and the much-denounced oil and mining companies had not developed the necessary technology for drilling that was then modified to deal with ice.[65]

Sustainability and Habitability: Distinguishing the Global from the Planetary

The difference between the global and the planetary is perhaps best illustrated by a quick contrast between two ideas central, respectively, to the two categories in question here, the globe and the planet. These are the ideas of sustainability and habitability.

Sustainability is a deeply political idea in the Arendtian sense of the word *politics*; it allows for the emergence of novelty in human affairs in a way that always involves some discussion about the welfare of the unborn. It owes its development to Europe's experience of agriculture and farming at a time of European expansion and thus belongs firmly to the history of the global.[66] The most widely used definition of sustainable development is the one that the World Commission on Environment and Development, often known as the Brundtland Commission after its chair Geo Brundtland, adopted in 1983 in its publication *Our Common Future*: "development that meets the needs of the present generation without compromising the ability of future generations to meet their own needs."[67] Paul Warde has written a differentiated history of the idea from the seventeenth century on — *Nachhaltigkeit* (the German word for lastingness or sustainability) is traceable in its earlier forms to the 1650s in texts on the management of agriculture and forestry in England, Germany, and France. His essay clarifies:

> The modern notion of sustainability largely [drew] on ideas developed in the late eighteenth and early nineteenth centuries when new understandings of soil science and agricultural practice combined to develop the idea of a *circulation* of essential nutrients within ecologies, and hence allow the perception that disruption to circulatory processes could lead to permanent degradation.[68]

One of the pioneers he mentions is Justus von Liebig, "chemist and admirer and follower of Alexander von Humboldt." Warde finds in Liebig's work "something like the modern conception of sustainability: that a society's development is beholden to fundamental biological and chemical processes [of the earth], but also that this was a complex dynamic system with feedback effects."[69]

Warde's statement makes visible how a certain incipient consciousness about earth processes—an incipient planetary consciousness, as it were—always lurks in the background whenever the question of sustaining human civilization is raised. But it *lurks in the background*: the idea of sustainability puts human concerns first. Donald Worster shows that the very idea of the earth as something finite belongs to a family of certain deeply anthropocentric ideas of which environment and sustainability are two important members. Worster describes William Vogt's *Road to Ruin* (1948) as "one of the first [texts] to use the word 'environment.'" Vogt defined *environment* as "the sum total of soil, water, plants, and animals on which *all humans depend*."[70] The word *environment* thus came to be something expressive of a human-centered concern, as if the only reason to speak of environing something was that the something was us. Fairfield Osborn's *Our Plundered Planet* (1948), published in the same year, was prepared to see the human species as "part of one great biological schema" while being sensitive to rich-poor differences. He was familiar with the facts of the deep history of the planet as they were understood in his time, but, like others, he had his sights firmly trained on what that history meant for humans. His aim was to help humans "learn to care for the greater good of nature and of humans as part of that whole," the idea of a "whole" referring in this case to issues like balance and harmony between humans and their earthly environment.[71]

This anthropocentric idea of sustainability dominated the twentieth century and continues beyond it as a mantra of green capitalism.[72] An absurd extreme of such a humanocentric conception was demonstrated early in the last century when the idea of "maximum sustainable yield," adapted from the history of "scientific" management of forests, became hegemonic in the literature on "managing fisheries." Peter Anthony Larkin put the matter with a touch of humor when he gave a keynote address to the Annual Meeting of the American Fisheries Society in 1976:

> About 30 years ago, when I was a graduate student, the idea of managing fisheries for maximum sustainable yield was just beginning to really catch on. . . . Briefly, the dogma was this: any species each year produces a harvestable surplus, and if you take that much, and no more, you can go on getting it forever and ever (Amen). . . . Moreover, it was assumed that the animals were well aware of what was being organized for them as their role in the scheme of things. Organisms were allowed to breed with those of their own species, or interact with individuals of other species, but not in ways that might upset the maximum sustained yield.[73]

In the literature on sustainability, earthly processes constitute a mute background for human activities. Stephen Morse's book on the subject of sustainability devotes only one of its 259 pages to the history of life on the planet, and that only because he needs to give the issue of sustainability an earthly context. But he points out that the word *sustainability* is not "used much" in describing life's continuity on this planet: "Instead we talk of the 'durability' or 'resilience' of life; its ability to continue after shocks and protuberances, of which there have been many since the birth of the planet." Now there, in that fragment of a sentence, a glimpse of a planetary consciousness shines through. But the word *sustainability*, as Morse correctly insists, applies only to humans. It is "a human-centric term," he acknowledges, and is "applied to people and the interactions we have with our environment. Thus, when we are talking of the role of biology within sustainability, we mean the role that biology plays vis-à-vis people, and we are talking of very short timescales relative to" those involved in the history of life.[74]

The key term in planetary thinking that one could contrapose to the idea of sustainability in global thought is *habitability*. Habitability does not reference humans. Its central concern is life — complex, multicellular life, in general — and what makes *that*, not humans alone, sustainable. What, ask ESS specialists, makes a planet friendly to complex life for hundreds of millions of years? The problem of habitability therefore should be distinguished from the discussion on life that has gone on in the humanities under the rubric of biopolitics. The idea of biopolitics that connects life to questions of disciplinary power, state, capitalism, and so on and rejects "a biological or metaphysical thematization of life" would squarely be a part of what I have characterized here as global thought.[75] The question at the center of the habitability problem is *not* what life is or how it is managed in the interest of power but rather what makes a planet friendly to the continuous existence of complex life.

Of course, the difficulty scientists face in discussing what makes a planet habitable is that the sample size of habitable planets available so far for study is only one. The necessary pluralism of the planetary thus appears to come somewhat undone with the question of life and habitability. But, as Langmuir and Broecker write, "While Earth's history is inevitably specific as a story of one planet, principles that it embodies [such as evolution by natural selection or 'increased stability through networks and increased access to and utilization of energy'] appear likely to apply on a universal scale."[76] The immediately relevant point is that humans are not central to the problem of habitability, but habitability is central to human existence. If the planet were not habitable for complex life, we simply would not be here. This is illustrated, for instance, by the share of oxygen in the atmosphere, which is currently around 21 percent and has been stable for a very long time.[77] As Langmuir and Broecker point out, this is "a striking disequilibrium state, because O_2 is such a highly reactive molecule." Oxygen reacts with "metals, carbon, sulfur, and other atoms to form oxides."[78] "What controls the atmospheric O_2 concentrations today?" ask Kump, Kasting, and Crane in their book on ESS:

> The answer, surprisingly, is that we do not know for sure, although researchers do have a number of ideas. Whatever the oxygen control mechanism is, it appears to be very efficient. The modern atmospheric O_2 level is 21% by volume, or 0.21 bar. It seems unlikely that the O_2 concentration has strayed from this level by more than ±50% since the last Denovian Period, about 360 million years ago. The evidence is that forests have existed since that time and, while they have always been able to burn, they have never disappeared entirely.[79]

An O_2 molecule resides in the atmosphere for four million years before getting absorbed into the earth's crust. "This may sound like a long time," remarks Lenton, "but it is far shorter than the 550 million years or so over which there have been oxygen-breathing animals on the planet. It is also far shorter than the 370 million years over which there have been forests." "Thus, remarkably," he concludes, "the amount of atmospheric oxygen has remained within habitable bounds for complex animal and plant life despite all of the oxygen molecules having been replaced over a hundred times."[80] This remarkable stability of the share of oxygen in the atmosphere allowing us to breathe is ensured by the Earth system or what I have called "the planet."

Earth system scientists appear to agree that different forms of life both on land and in the sea, the rate of burial of organic carbon in the sea, and the phosphorus and long-term carbon cycles of the planet all have a role to play in replenishing and maintaining the share of oxygen in the atmosphere that allows complex life to flourish.[81] This is why within a planetary mode of thinking, the threat of the Anthropocene lies in what it might mean not simply for immediate human futures but for long-term futures as well. Global warming produces for Earth system scientists the fear of another great extinction of life—possible in the next three hundred to six hundred years—that might make the planet regress to a more primitive level of biodiversity.[82] As Langmuir and Broecker argue, fossil fuel, soil, and biodiversity are critical to human flourishing, and they have two things in common: they all have to do with the history of life on the planet, and none of them are renewable on human scales of time.[83] The planetary, ultimately, is about how some very long-term planetary processes involving both the living and the nonliving have provided, and keep providing, the enabling conditions for both human existence and flourishing. Our recent interference with some of these processes, however, has raised for humans a particularly intractable question with a sense of urgency surrounding it, the question—to use the evocative words of William Connolly—of "facing the planetary."[84]

Facing the Planetary

For all their differences, thinking globally and thinking in a planetary mode are not either/or questions for humans. The planetary now bears down on our everyday consciousness precisely because the accentuation of the global in the last seventy or so years—all that is summed up in the expression "the great acceleration"—has opened up for humanist intellectuals the domain of the planetary. As discussed before, even the everyday distinction we make between renewable and nonrenewable sources of energy makes a constant reference, by implication, to human and geological scales of time, to the hundreds of millions of years that the planet would take to renew fossil fuels. Similarly, all talk about there being "excess" carbon dioxide in the atmosphere refers implicitly to the normal rate at which the carbon sinks of the planet take up this gas. Langmuir and Broecker emphasize the critical importance to humans of counting soils and biodiversity among the "nonrenewable resources," not simply fossil fuels.[85] Practical plans to make profit by developing technology that uses the sun as an infinite source for energy for industrial and industrializing societies are attempts to bring into the fold of the global an aspect of what we have called the planetary. We are all living, whether we acknowledge it or not, at the cusp of the global and the planetary. The age of the global as such is ending. And yet the quotidian is about both invoking the planetary and losing sight of it the next moment.

Is this forgetting a problem? Connolly has asked this question. "By 'the planetary,'" he writes,

> I mean a series of temporal force fields, such as climate patterns, drought zones, the ocean conveyor system, species evolution, glacier flows, and hurricanes that exhibit self-organizing capacities to varying degrees and that impinge upon each other and human life in numerous ways. . . . The combination of capitalist processes and the amplifiers in nonhuman geological forces must be encountered together. Such a combination poses existential issues today.[86]

Connolly is right to say that "the combination of capitalist processes" and the planetary ones have to be "encountered together." But what does it mean to encounter them "together?" How exactly does one encounter *together* (in thought) disparate forms of thinking even when the phenomena they refer to appear intertwined and when the global and the planetary—with their respective anthropocentric and nonanthropocentric emphases and with their references to vastly different and incommensurable scales of time—often represent two rather different orientations to this entity on and from which we live?

The global, as I have said, refers to matters that happen within human horizons of time—the multiple horizons of existential, intergenerational, and historical time—though the processes might involve planetary scales of space. Planetary processes, including the ones that humans have interfered with, operate on various timetables, some compatible with human times, others vastly larger than what is involved in human calculation. Thus, air and surface water have "short recycling times," as do many metals, but soils and ground water take "'thousands of years'" to replenish themselves. "Biodiversity," writes Langmuir and Broecker, "is perhaps the most precious planetary resource, for which the timescale of replenishment, known from past mass extinctions, is tens of millions of years."[87] Humans today have become a planetary force in that they can interfere with some of these very long-term processes, but "fixing them" with the help of technology is still well beyond our present capabilities. What would it mean for us to bring together in our thought all these different timescales and, in Connolly's terms, face them?

Temporality, however, is not the only thing that distinguishes the global from the planetary. The two modes of thinking represent two different kinds of knowledge and, for humans, two different ways of comporting themselves to the world within which they find themselves.[88] The global with humans at its center is ultimately all about forms and values. This is why the planet when equated to the globe can be politicized (we can talk about its deliberate destruction by Exxon or about creating "planetary sovereignty").[89] Debates on issues like climate justice, climate refugees and their rights, democracy and global warming, climate change and inequalities of income, race, gender, and the good and bad Anthropocene proceed on the assumption that we have ideas, however contested by competing ideas, about ideal *forms* of justice, rights, democracy, and so on in order to be able to judge and pronounce on a situation. These questions that deeply involve the question of forms and the politics of debating them belong to the global.

But the planetary as such, disclosing vast processes of unhuman dimensions, cannot be grasped by recourse to any ideal form. There is no ideal form for the earth as a planet or of its history or for the history of any other planet. While the planetary mode of thinking asks questions of habitability, and habitability refers to some of the key conditions enabling the existence for various life-forms including *Homo sapiens*, there is nothing in the history of the planet that can claim the status of a moral imperative. It is only as humans that we emphasize the last five hundred million years of the planet's life—the last one-eighth of the earth's age—for that is the period when the Cambrian explosion of life-forms occurred, creating conditions without which humans would not have been. From the viewpoint of anaerobic bacteria, however, which lived on the surface of the planet before the great oxygenation of the atmosphere about 2.45 billion years ago, the atmosphere might look like a history of disasters (as recognized by such human-given names as the Oxygen Holocaust). The planet exists, as Quentin Meillassoux says, "as anterior to the emergence of thought and even of life—*posited, that is, as anterior to every form of human relation to the world.*"[90]

The Planet and the Political

Faced with the radical otherness of the planet, however, a deeply phenomenological urge on the part of many scientists is to recoil back into the human-historical time of the present and address the planet as a matter of profound human concern—as a critical question of human futures and as an entity to be governed by humans. But the governance question, whether posed in terms of sustainability or habitability, is at base an existential concern that can only belong to the present. The critical difference is that in answering this existential question, Earth system scientists' ideas point to a profound shift in conceptions of how humans are to dwell on Earth. It is as if Schmitt's land/sea opposition, the opposition between our "terrestrial modes of being [*eines terranen Daseins*]"—signifying the desire for rest, stability, house, property, marriage, family, and so on—and our "maritime existence"—symbolized by the restless and perpetual movement of the technology-driven, imperial-European, oceangoing ship—has finally come to be realized in the picture of a geoengineered, "intelligent" planet making its voyage through the infinite seas of the universe.[91]

In 1999 Hans Joachim Schellnhuber, the physicist who set up the Potsdam Institute for Climate Impact Research in 1992, asked what Erle Ellis regards as "the pivotal question" of the Anthropocene: "'Why should Prometheus not hasten to Gaia's assistance?' . . . Can humans help to bend Earth's trajectory towards better outcomes for both humanity and non-human nature?"[92] Ellis endorses the view, albeit cautiously: "Hopes for a technocratic Prometheus are more than just pipe dreams.... The prospects for anthropocenes much better than the one we are now creating are very real."[93] Lenton writes, "Whilst human transformation of the planet was initially unwitting, now we are increasingly collectively aware of it.... This changes the Earth system fundamentally, because it means that one species can consciously, collectively shape the future trajectory of our planet." Such evolving "human consciousness" itself becomes a "new property of the Earth system."[94] "Human civilization," we hear from Langmuir and Broecker, "has led to the first global community of a single species, destruction of billions of years of accumulation of resources, a change in atmospheric composition, a fourth planetary energy revolution, and a mass extinction." Yet, they argue, "there is the potential in human civilization for Earth to pass from 'habitable planet' to 'inhabited planet,' i.e., one that carries intelligence and consciousness on a global scale, for the benefit and further development of the planet and all its life."[95]

This human concern opens out into another argument that is truly planetary and yet is drawn back immediately into human horizons. How long can a highly developed technological civilization last, ask Langmuir and Broecker? "*Does such a civilization self-destruct in a few hundred years or last for millions of years? For such a civilization to last, the species driving the technology must [consciously and collectively] sustain and foster planetary habitability rather than ravage planetary resources.*"[96] Hence, their hope that humans would be able one day "to view themselves and act as an integral and responsible part of a planetary system."[97] This, they write in concluding their book on the history of the Earth system, "is the challenge of human civilization, to become a part of a natural system to permit and perhaps even to participate in further planetary evolution."[98]

Lenton and Latour—even as they acknowledge that "in politics the blind lead the blind"—express the view that hope might lie in scientists collaborating with "citizens, activists, and politicians" so that enough "sensors" (the scientific-technological equivalent of the blind person's white cane) could be put in place to enable them all "to *quickly* realize [and presumably fix] where things are going wrong." Being thus able to track "the lag time between environmental changes and reactions of societies," they add, "is the only practical way in which we can *hope* to add some self-awareness to Gaia's self-regulation."[99] As a student of human pasts and politics, I find this vision of a future where scientists, activists, and politicians and their respective constituencies move "quickly" to recognize errors made on a very large social scale certainly reasonable but perhaps unlikely.

In any case, the language of hope (and despair), when we are confronted with the planet, turns us toward the present, for hoping and despairing are things we do in the human *now* while the planetary places humans against an unhuman backdrop. This seeming rapprochement between the timescale of the planetary and the time in which human hope and despair arise is intellectually fragile. It remains open to criticism for its assumption that humans can somehow get around being the kind of "pluriverse" that they are and that Schmitt saw as the ground for the friend/enemy distinction in his famous concept of the political.[100] The human political, one could say following Schmitt, is constitutionally plural and, as we know from problems of the IPCC trying to produce strategies for governing climate change, it cannot be easily subordinated by any one rational strategy. The anthropocenic regime of historicity as visible in ESS sets humans against a background of relationships and time that necessarily cannot be addressed from within the temporal horizon of human experiences and expectations—that is, from within the global regime of historicity. Yet that is the reconciliation that even Earth system scientists seek to achieve as historians of human futurity. Their understandably human and presentist concerns end up obscuring the profound otherness of the planet that their research also reveals.

The hope that humans will one day develop technology that will remain in a commensalist or congruent relationship with the biosphere for a period stretching into geological timescales—such a hope belongs to the realms of a reasonable utopia.[101] In spirit, it is no different from what Felix Guattari once wrote in his climate-unaware *Three Ecologies* (1989). With a sense of prophecy that today—after all the debate about geoengineering and humans as the "God species"—must at least sound a little dubious, Guattari wrote that "the health" of the planet earth

> will be increasingly reliant on human intervention, and a time will come when vast programmes will need to be set up in order to regulate the relationship between oxygen, ozone and carbon dioxide in the Earth's atmosphere. . . . In the future much more than the simple defence of nature will be required; we will have to launch an initiative if we are to repair the Amazonian "lung," for example.[102]

The "Amazonian 'lung,'" like the conveyor belt (the North Atlantic Meridional Overturning Circulation) of the Atlantic, may well be a part of the Earth system, and it is much easier perhaps for humans—in human time—to destroy than to fix such parts. To try to derive any ethical or moral lessons from our new understanding of the Earth system—the multiple networks of connections in which our bodies are like nodal points, simply a site that many connections pass through—is to try to bring within the grasp of the global (the domain of forms and values and therefore of the political) the planetary that not only out-scales the human but also, as I have said, has nothing moral or ethical or normative about it. This urge itself is symptomatic of the predicament that the Anthropocene is. It arises from the realization that the reach of the global, something Guattari called Integrated World Capitalism, has through the intensification of its energies completely discredited the nature/society or subject (human)/object (nature) distinction that has been taken for granted for so long in all discussions of modernity.[103] More than that, the institutions of human civilization, including technology, have interfered with some critical planetary processes. Planetary climate change is precisely an example of this point; humans have broken the planet's short-term carbon cycle by producing an excess amount of carbon dioxide that human institutions and technology cannot yet manage to recycle.

Facing the planetary then requires us to acknowledge that the communicative setup within which humans saw themselves as naturally situated through categories like earth, world, and globe has now broken down, at least partially. Many traditions of thought, including some religious ones, may have considered the earth-human relationship special; with regard to the planet, though, we are no more special than other forms of life. The planet puts us in the same position as any other creature.[104] Our creaturely life, collectively considered, is our competitive animal life as a species, a life that, *pace* Kant, humans cannot ever altogether escape.[105] The point was tragically illustrated during the devastating fires that Australia suffered at the end of 2019 and the beginning of 2020 when the department of environment of the government of South Australia took the decision to "destroy"—"in accordance with the highest standard of animal welfare"—up to ten thousand feral camels because the animals were competing directly with rural Indigenous communities for "scarce food and drinking water."[106] Humans and camels in this story are simply two earthly creatures competing for the same resources. Our encounter with the planet in humanist thought thus opens up a conceptual space for the emergence of a possible philosophical anthropology that will be able to think capitalism and our species life together from *both* within and against our immediate human concerns and aspirations.

Political thought since the seventeenth century has been grounded in the idea of securing human life and property. This thought has remained constitutionally indifferent to human numbers—as it was after all the human individual who was the bearer of life, the possessor of rights, and, finally, the recipient of welfare. This indifference to total number of humans translated into an indifference to the biosphere, the reigning assumption being that the globe was always resourced enough to support in perpetuity the human-political project no matter how demanding humans became of the earth. But our encounter with the planet or the Earth system allows us to see how some of the basic assumptions of this tradition now stand challenged. The harder we "work" the earth in pursuit of the worldly flourishing of a great number of humans, the more we encounter the planet. If human institutions, technology, and profit seeking that have so far worked in tandem to "secure" human life expanded to a point whereby planetary cycles broke down, the seas got warmer and more acidic, forests vanished, biodiversity was stressed and species extinction hastened, the number of refugees in the world (now calculated to be around sixty-five million) likely trebled, the frequency of "extreme weather" events increased, and the labor of humans and animals got displaced by the work of artificial intelligence, then a profound and tragic irony would reveal itself in such a course of human history. The institutions humans have used so far to secure human life have reached a point of expansion and development whereby that very fundamental premise of human politics—securing human life—is undermined. Late capitalism, in this sense, destroys the human-political project the world over. In such circumstances, there is surely the danger, as Latour points out, of a rebarbarization of the world, a prospect that many authoritarian leaders and parties today implicitly or explicitly embody and hold out.[107]

If the climate crisis of human flourishing brings into view planetary processes that humans in the past simply ignored, bracketed, or took for granted, it is reasonable to ask for an ethic that allows humans to develop "everyday tactics for cultivating an ability to discern the vitality of matter."[108] But we also have to agree with Jane Bennett that such "attentiveness to matter and its powers will not solve the problem of human exploitation or oppression. . . . It can [only] inspire a greater sense of the extent to which all bodies are kin in the sense of [being] inextricably enmeshed in a dense network of relations."[109] Posthumanism by itself cannot address the political. Any theory of politics adequate to the planetary crisis humans face today would have to begin from the same old premise of securing human life but now ground itself in a new philosophical anthropology, that is, in a new understanding of the changing place of humans in the web of life and in the connected but different histories of the globe and the planet.

As the geologist Jan Zalasiewicz once observed, "It is hard, as humans, to have a perspective on the human race."[110] What indeed are the perspectives that ESS offers? Augustine turned to writing his *Con-*

fessions when he realized that he had become a "question" for himself.[111] We could similarly ask, If one reads ESS as providing an (auto)biography of humans when humans have become a question for themselves, what indeed is that question that motivates this narrative? The question itself remains unasked, but many second-order, derivative questions swim around in its gravitational field. Are humans now a "God species?" Should humans make kin with other nonhuman beings? Should human societies aim to become a part of the natural systems of the planet? Will the earth become an "intelligent" planet thanks to the integration of the technosphere and the biosphere? Such questions — not yet answerable yet gaining in force everyday — mark out how the category *planet* enters humanist thought, as a matter of human-existential concern, even as we come to realize that the planet does not address us in quite the same way as our older categories of *earth*, *world*, and *globe*. We will return to these questions toward the end of this book. The next part of the book, however, explores how this awakening to the scale of the planetary makes us rethink certain key themes in the global history of modernity and modernization.

PART II

*

The Difficulty of Being Modern

4
The Difficulty of Being Modern

There is an important part of climate change discourse that, it may be said, sees itself as a continuation of the critique of the inequities of globalization and is therefore quite compatible with Schmitt's schema of appropriation→distribution→production we discussed in the last chapter. This is the literature on "climate-justice" issues. But we need to modify the Schmittian schema in one important respect: with the warming seas and their rising levels, with increasing droughts and superstorms, and with refugee numbers swelled directly or indirectly by climate change, the struggle today is not just about distribution or justice, it is about appropriation as well, a subject that directly addresses security studies and international relations, touching on fundamental political questions about sovereignty. I could cite many examples to illustrate this point, but let me just quote Phillip Muller, the then ambassador of the Marshall Islands to the United Nations, speaking to Columbia University's newly set up Center for Climate Change and Law in the year 2009:

> The seas are rising, and some decade—no one knows which country of the twenty-nine coral atolls and five islands, located midway between Hawaii and Australia, is going to be under water. When that happens, a number of novel legal questions will arise. If a country is under water, is it still a state? Does it still have a seat at the United Nations? What becomes of its exclusive economic zone, and the fishing rights on which it depends for much of its livelihood? What countries will take its displaced people and what rights will they have when they arrive? Do they have any recourse against those states whose greenhouse gas emissions caused this plight?[1]

In this quote, the impact of climate change raises all the issues that marked the Schmittian schema—of sovereignty and justice (distribution), production (fishing rights), and appropriation (loss of land, exclusive economic zones, refugees turning up elsewhere). The problem of justice is formulated here in political terms that belong to the history of globalization: would the nations and peoples suffering the impact of climate change have "any recourse," as the Ambassador put it, "against those states whose . . . emissions caused" their plight? At the heart of the climate problem, the justice question introduces the matter of "uneven development." An anthropocentric concern all right, but one that is directly connected to debates about capitalist development and world markets.

But there is yet another concern of developing nations that underlies their complaints about the inequities of the impact of climate change and one that I consider crucial to the argument about the relationship between climate and global capital: this is the widespread desire for growth, modernization, development, whatever one calls it, in the less developed nations of the world. The question of development—in fact, the right to development—was at the center of the so-called climate-justice debate that was initiated in 1991—a year after the first report of the IPCC was published—by the Indian environmental activists Anil Agarwal and Sunita Narain, whom we have met before. To my knowledge, they were the first to propose that the national emissions of greenhouses gases (GHGs) be computed on a per capita basis. Agarwal and Narain objected to sweeping use of the word *human*—their immediate target a report of the World Resources Institute (WRI) on the "global environment"—and what they saw as the spurious "one world–ism" of the West.[2] Agarwal and Narain saw all this as an "excellent example of environmental colonialism" that, they suspected, actually "intended" to "perpetuate the global inequality in the use of the earth's environment and its resources" by blaming 'developing countries' for global warming" when "the accumulation in the earth's atmosphere of these gases [GHGs] is mainly the result of the gargantuan consumption of the developed countries, particularly the United States."[3]

For Agarwal and Narain, it was as though climate change was ushering in a cruel and unfair "regime of historicity" that threatened to shut down the future that India and China saw themselves as pursuing as they became two independent nations in the late 1940s and more vigorously since the 1980s: an open vista of modernization that the United States and Soviet Union inspired after the Second World War.[4]

> Many developing countries fear that the proposed climate convention [Rio 1992] will put serious brakes on their development by limiting their ability to produce energy, particularly from coal . . . , and under-

take rice agriculture and animal care programmes. . . . The focus [in the West] today is on poor developing countries[,] and their miniscule resource use is frowned upon as hysteria is built up about their potential increase in consumption. . . . The dream of every Chinese to own a refrigerator is being described as a curse.[5]

Thus, the argument that came to be known as "climate justice" could also be seen as a strategy for bargaining, in effect, for a longer life for a developmental regime of historical time for nations like India and China (which is not to deny their point about climate justice).

One cannot debate the politics of climate change without looking at how issues of "development" affect subaltern modernizers of history. Take the simple question of the market for air conditioning in India. On October 12, 2016, negotiators from 170 nations met in Kigali, Rwanda, and agreed to phase out the use of heat-trapping hydrofluorocarbons (HFCs) used in making the cheapest air conditioners that aspirational families, often low in the social hierarchy, have begun to buy in countries like India. The air conditioners enable them to deal with summers that get hotter with every passing year. HFCs trap heat one thousand times more effectively than carbon dioxide.[6]

Economist Michael Greenstone reports in the *New York Times* that while 87 percent of US households have air conditioning, the figure for India is 5 percent (or 6–9 percent according to some others). Annually, Delhi currently gets five or six days when the average temperature goes above 95°F; by the end of the century the number of such days is expected to rise to seventy-five. The mortality effects of each additional day over 95°F "are 25 times greater in India than in the United States, where the use of air-conditioners reduced by 80 percent the number of heat-related deaths between 1960 and 2004."[7] In another *New York Times* article, Ellen Barry and Carol Davenport report that scientists claim "a surge in the use of HFC-fueled air-conditioners would alone contribute to nearly a full degree Fahrenheit of atmospheric warming over the coming century—in an environment where just three degrees of warming could be enough to tip the planet into an irreversible future of rising sea levels, more powerful storms and deluges, extreme drought, food shortages and other devastating impacts." Yet this "surge" is exactly what is happening in India, where, according to the same report, "the purchase of a first unit—not a second or a third—is driving growth." "Every time government salaries are raised," Barry and Davenport write, paraphrasing an Indian official, "air-conditioner purchases surge," even among urban working-class families.[8]

Barry and Davenport's report captures something of what we may call, following Ranajit Guha, the "small voices" of contemporary history, the voices of those who have to deal with a warming world while expressing and pursuing their aspiration to social mobility and modernization.[9] It is also necessary to keep in mind the subject of population growth in India, especially in the cities. Globally, 50 percent of all the growth in human population between now and 2100 is supposed to come from eight countries, of which India and Pakistan are two (the others are all in Africa: Nigeria, Tanzania, Democratic Republic of Congo, Niger, Uganda, and Ethiopia).[10] While on certain aspects of the population question, aspirations and gender-justice can indeed be brought in line with the larger task of democratically reducing population through development—by ensuring women's access to education, job opportunities, and contraception—there still remains the problem of the rapid development of megacities, a world that Mike Davis appropriately christened a "planet of slums."[11] The overall population in India between 2001 and 2011 grew by about 17 or 18 percent. The city of Bangalore grew "a whopping 47 per cent, its density growing from 2,985 people per square kilometer in 1991 to 4,378 in 2010. Delhi grew by 21 per cent between 2001 and 2011."[12]

It is not surprising then to read that "a thrill goes down Lane 12, C Block, Kamalpur [Delhi] every time another working-class family brings home its first air-conditioner. Switched on for a few hours, usually to cool a room where the whole family sleeps, it transforms life in this suffocating concrete labyrinth where the heat reaches 117 degrees in May." "You wake up totally fresh," says Kaushilya Devi, a housewife. Her husband bought a unit last May. "I wouldn't say we are middle class," she adds, "but we are closer." A bank manager, S. S. Pathak, is grateful that the air-conditioner enabled his children to study for their medical school entrance examination—they could now "manage late-night study sessions without nodding off or being devoured by disease-carrying mosquitoes." Another interviewee, Sandhya Chauhan, and her family "live in two musty, windowless subterranean rooms, which turn stifling on summer nights, leaving six sweat-soaked adults to fidget, toss and pace until the morning":

> But it was never as awful as this May [2016], when the temperature crept so high that Mrs. Chauhan's friends speculated that the earth was colliding with the sun.... After a doctor warned Mrs. Chauhan that heat exhaustion was affecting their oldest son's health, her husband bought an air-conditioner on credit.... The purchase has changed the way they see themselves.... "Education is teaching people to take care

of themselves," she said. "Now that we are used to air-conditioners, we will never go back."¹³

These gendered, subaltern, aspirational voices make it clear that our sense of ordinary human flourishing and even of democracy in a warming world depends on making available to all energy that is cheap and plentiful. Arjun Appadurai's insightful words on such everyday aspirations bear repetition:

> Aspirations to the good life are part of some sort of system of ideas . . . that locates them in a larger map of local ideas and beliefs about . . . life and death, the nature of worldly possessions, the significance of material assets over social relations, the relative illusion of social permanence for a society, the value of peace and warfare . . . local ideas about marriage, work, leisure, convenience, respectability, friendship, health, and virtue.¹⁴

Yet imagine the future that Kaushilya Devi and Sandhya Chauhan face as nations make the decision to switch over to alternatives for HFCs. The replacements, says Stephen Yurek, president of the Air-Conditioning, Heating and Refrigeration Institute, are "more flammable and toxic" and hence need better-designed and more expensive air-conditioning units and better-trained workers to install them. India has understandably asked for a slow transition: to delay the elimination of HFCs until 2031 and to phase it down to about 15 percent of 2029 levels by 2050 provided there is some aid forthcoming from the developed countries whose experts say that it is crucial to ban HFC before the air-conditioning boom happens. In China, only 5 percent of urban residents had air conditioning in the 1990s; in ten years the figure rose to 100 percent.¹⁵ Greenstone comments on the obvious irony of the situation: "The very technology that can help to protect people from climate change also accelerates the rate of climate change." But for now, "India is heavily focused on current residents who face risks that simply don't exist in wealthy countries like the United States."¹⁶ Whoever is prime minister of India in the coming decades will need the consent of the Kaushilya Devis and the Sandhya Chauhans of the country to fulfill India's international obligations on HFCs.

The Posthuman and the Postcolonial

How do we square the reality of these popular aspirations that play out over electoral cycles and institutional politics with what scholarly voices from Earth System Science and from what we gather under the rubric "posthumanism" tell us about an entangled world, distributed agencies, the role of planetary processes, the nonhuman, and so on? The ambassador of the Marshall Islands, whom I quoted earlier, may indeed speak of the islanders' right to fish tuna in their exclusive economic zone of the sea, but think of the role of the tuna! The tuna, following changing gradients of oceanic temperatures, could very well decide to turn up in other waters more friendly to the logic of their habitation and reproduction.

A nonanthropocentric view of the world, as discussed in the previous chapter, is integral to Earth System Science, and therefore whether we speak of "capitalism in the web of life" or of the "Capitalocene," it is difficult if not impossible to ignore—when considering the issue of climate change—the question of the agency of the nonhuman and the nonliving. It is not surprising at all that the planetary crisis of climate change should invite comment from those who broadly write under the rubric of posthumanism—Bruno Latour, Donna Haraway, Anna T. Sing, Jane Bennett, Rosi Braidotti, and others. Elizabeth Povinelli's *Geontologies*, Déborah Danowski's and Viveiros de Castro's *The Ends of the World*, the political theorist William Connolly's *Facing the Planetary*, Michael Northcott's *A Political Theology of Climate Change* are attempts to generate a grammar of a new politics combining the agencies of humans and nonhumans.[17] The epistemological appeal of this move toward a posthuman description of the world—and the desire to create a corresponding sense of the political (think of Latour's idea of the "parliament of things")—is very well expressed indeed by Jane Bennett in her book *Vibrant Matter*, where she describes the nature/culture distinction as giving us not so much a wrong as a "thin" description of the world. Posthuman studies—her book suggests, using creatively this Geertzian opposition between thick and thin—provide the much needed corrective of "thick" description: "Theories of democracy that assume a world of active subjects and passive objects begin to appear as thin descriptions at a time when the interactions between human, viral, animal, and technological bodies are becoming more and more intense."[18]

Even if we conceded that views that look on agency as something distributed between humans and nonhumans do perhaps give us better descriptions of how the planet and life on it actually work, a critical question would still remain: Why do modern humans, in spite of this knowledge, remain more attached to the nature/culture distinction, that is, to what Bennett calls a "thin description" of reality? How does one account for the desire for modernity or so-called development—or at least for the conveniences of modernization—among many if not most humans everywhere? What is the relationship between the projects for modernization that were initiated in the third world by anticolonial modernizers of formerly colonized or "new" nations of the 1950s and 1960s in Asia, Africa, the Pacific, and elsewhere, the desire for capitalist growth and progress in populous nations like India and China, and the climate crisis today?

The existing debate in the human sciences on climate change—even when it acknowledges (and it mostly does) the reasonableness of the "climate-justice" position—gives us no insight into the history of these third-world desires, why and how and through what kind of intellectual and social history development and progress came to be such valued notions in India, China, postcolonial Egypt, Indonesia, or Papua New Guinea. And not only desires. Technological domination of nature was experienced as masculinity far beyond the boundaries of the so-called West. Even in my own economically depressed history of the colonial Bengali-Hindu middle class, a talented, young, and later well-known poet Premendra Mitra (1904–1988), intoxicated by the seemingly triumphant success of the labor of "Western humanity"—Arendt's *animal laborans*—and taking it to represent the pinnacle of the history of human labor as a whole (notice the use of the word *lazily* in the poem)—thus exulted in human ravaging of the earth by portraying, in an entirely masculine manner, the earth itself as wanting to be so ravaged:

> The earth begs for the thrust of the plough
> The ocean for the helm.
> Metals, imprisoned in the palace of the Deep,
> Pine away for [the touch of] man.
> The boisterous river wants to fall into chains,
> Into bondage to the bridge.
> No time, alas, to gaze
> Lazily on the beauty of the world.[19]

Marxist critics who locate the roots of global warming in the story of global capitalism want to rename the Anthropocene and call it the Capitalocene or something else that alludes to its social genesis. But they are silent on the question of how or why visions of modernized futures came to seize the imagination of the middle and other classes of nations that were once colonies of European powers. If there is any agency of concrete humans in the Marxist literature on Capitalocene—that is, agency in excess of what may be attributed to the abstract logic of capital—it belongs to industrial captains and elites in boardrooms and governments who make economic decisions and not to the elite, middle, or subaltern classes of Asia and Africa.[20] In his Keynote lecture to the Millennium Conference of 2015, Bruno Latour explained humanity's willingness to pay this epistemological "price" (the nature/culture distinction) by referring to its practical "advantages": "Of course, this price is worth paying in many situations. Great progress is made by those who localise parts, add relations, build mechanisms, link elements with cause and effect relations, and build a scale model of the whole set up. The advantage of such a procedure is not in question."[21]

* * *

The nature/culture separation, or what amounts to an ontological separation of the human from the nonhuman, results, as Latour points out in his classic *We Have Never Been Modern*, in certain projects of "purification." Modernity and the capitalist mode of production are indeed unthinkable with their accompanying processes, both intellectual and practical, of extracting out of "nature" various entities in their supposed states of purity. This is what Jane Bennett refers to as working with a "thin description" of nature. Think of a commodity as elementary as "land." When a piece of land is sold, it is sold as a piece of abstraction, a two-dimensional figure on a map devoid of, say, all the forms of life that inhabit it except maybe those of immediate monetary value to humans. Or think of metals and minerals. Seldom do they occur in nature in a pure form. And it is no wonder that petroleum plants have "distilleries"—the very name says it all. As Zalasiewicz and his colleagues say, pure or alloyed metals were

> rare on prehuman earth, where gold and (less commonly) copper and iron were found to naturally occur in amounts that could be exploited. Commencing only in the Holocene ... humans have isolated metals by smelting from their compounds, beginning with lead, silver, and tin (most copper and iron, too, had to be extracted from compound ores). In a burst of innovation from the late 18th to mid-20th century, most metals were isolated, including some never known to have existed previously in native form, such as magnesium, calcium, sodium, vanadium, and molybdenum and some that only occur rarely and in miniscule amounts, such as aluminium, titanium, and zinc.[22]

Novel metal alloys include bronze, brass, pewter, and iron-carbon alloys, "often with chromium, molybdenum, and other metals." And this production of purity has also led to the proliferation of what Latour calls "hybrids." Humans have now produced a "wide range of synthetic minerals ... novel forms of garnet ... [and] crystalline materials" for use in lasers, such as boron nitride (Borazon), "an industrial abrasive." Boron carbide is another such hybrid metal that is used in tank armor and bulletproof vests, "while tungsten carbide is used as the balls in ballpoint pens." The Inorganic Crystal Structure Database lists, write Zalasiewicz and his colleagues, "more than 180,000 different types of 'synthetic' mineral-like compounds" made by humans.[23] One could also add to this list "Novel Human-Made Minerals" (at first, "human-mediated minerals") often associated with mining (made "by weathering of mineral slags, crystallization from mine drainage systems, or precipitation on tunnel walls as well as corrosion products around archaeological artifacts") and synthetic mineral-like compounds such as "mass produced ubiquitous building materials such as Portland cement" (the basis of concrete) and "clay-fired products such as porcelain and bricks." Less voluminous but equally widespread are "technological crystals, including those used in semiconductor devices, magnets, phosphors and other electronic applications."[24] Latour is absolutely right: the project of purification goes hand in hand with the proliferation of hybrids, a process that, as Latour argues, ultimately undercuts the very nature/society or nature/culture opposition that makes the projects of "thinning" nature or producing entities in a "pure" state possible at all.

All this is granted. But consider also this important point: if the desire for modernization/development of the vast non-Western middle classes were only a matter of utility, practical advantage, greed, or profit, this desire would simply seem crass and morally indefensible. One could then repeat with confident moral anger the aphorism ascribed to Gandhi that while there is enough in the world to fulfill everyone's needs, there was never enough to fulfill everyone's greed—and be done with critiquing modernization. If that were all there was to development and modernization, thinkers such as Amartya Sen (and Martha Nussbaum and others) would not have been able to build the famous "capabilities approach" to the problem or describe "development as freedom."[25] One needs to understand the ethical aspects of such desire if one is to plumb the depths of the human predicament today.

This is where, I suggest, the story of anticolonial, third-world modernizers has to be taken into account. Latour's engagement with the Anthropocene, for instance, has been grounded in his earlier critiques of what he memorably called "the Constitution of the Modern," a peculiar constitution that, thanks to its absolute separation (let's say, from the seventeenth century on) of nature from society, a version of the nature/culture opposition, allowed the proliferation of a multitude of hybrids (things that were neither purely natural nor purely social) while denying the actual work of translation between the two poles that brought the hybrids into being and insisting that the hybrids were a mere mixture—a mediation—of two separate and pure forms.[26] It is not difficult to see that the target of his criticism was clearly an entity he called "the West," "the Occident," "Western society," and the arrogant schema of its nature-society separation that helped it to dominate what was outside of it and also its own population by fabricating the themes of modernity and modernization.

Latour suggests, through some cryptic remarks, that this West—both the fabricator and a fabrication of the Modern—is not without history. What we have, however, are some very short, brilliant, and suggestive formulations, such as the one arguing that the Constitution of the Modern has become burdened with its own contradictions. It is not difficult either to put some rough bookends to the story of the Constitution of the Modern. Such bookends become visible from Latour's narrative: starting from the time of the Boyle-Hobbes controversy (as reported by Shapin and Schaeffer) in the seventeenth century and running through to our present, the moderns have scaled up the production of hybrids—of nature and culture—to such a degree that the constitution dependent on the maintenance of this distinction is at a point of collapse. Climate change confirms the depth of this crisis. Of course, there are those who become the subjects of the modern's constitution—both in the colonies and in Europe. They tell us human history is much, much older than this constitution and also that, apart from the aspect of scale, none really has ever been modern, surely not those proclaiming their modernity from rooftops. Latour's project not only divests from any kind of Eurocentrism, it also divests from the claim that this Constitution of the Modern describes how the world actually works, the actual networks of entanglement that he tries to make visible in his magnum opus, *An Inquiry into Modes of Existence*.[27]

Latour's project offers in many ways a profound critique of the world that the modern constitution has made possible. He proceeds by critiquing the nature/society opposition at the heart of this constitution and thus attempts to usher in a new world order—a parliament of things (hinted at in *The Pasteurization of France*, somewhat developed in *We Have Never Been Modern*, and fully presented in *The Politics of Nature*).[28] When Latour engaged with the Anthropocene and climate change—at least in the first drafts of his Edinburgh lectures that he generously made public and shared with friends and colleagues—his canvas expanded to take in James Lovelock's Gaia hypothesis, which he dexterously maneuvered to bring up the question of religion. This was completely legitimate—after all, Gaia was herself a religious figure. Latour staged a war between the people of Gaia who did not want to live by the Constitution of the Modern and those who did (the people of science). His thoughts went back to the period of "early modernity" by weaving his work through Hume's work on "natural religion." But the imaginary population of people who lived by "science" harked back to many of the themes familiar to Latour's readers. Critiquing the Constitution of the Modern is a project in favor of a more equal and substantially—and not just formally—democratic world.[29]

I completely agree with Philippe Descola's remark that "all in all," Latour's argument is "very convincing."[30] But where are the anticolonial, late-modern, and the late-modernizing leaders of Asia and Africa—the Nehrus, the Nassers, the Sukarnos, the Nyereres, the Senghors, the Frantz Fanons—in this story? Latour's argument in *We Have Never Been Modern* and elsewhere remains founded on a face-off between "we moderns from the Western world" or "the Westerners [and] the Whites (whatever nickname one might wish to give them)" on the one hand and the Indigenous peoples of America on the other, especially as represented in Phillipe Descola's ethnography of the Achuar people living on the border of Ecuador and Peru.[31] Are we to assume that anticolonial leaders desiring to "catch up" with the West—a desire that still propels the politics of India and China (remember Deng Xiaoping's "four modernizations" campaign?)—were simply advocating pale, unoriginal copies of their forerunners in the West—mimic, derivative desires condemned by history to repeat the West's folly—so that critiquing European modernizers takes care of their cases as well? Latour does not discuss debates on modernity that have obsessed postcolonial critics, from Anthony Appiah to Homi Bhabha. What concerns him more is how the project of modernization is doomed to failure. In the sixth Edinburgh lecture on Gaia, he remarked, "If you can still dispute whether 'we have never been modern' or not, who now disputes that 'we' will never be able to modernize the earth for lack of the five planets (according to calculations by 'global hectares') that would be needed to push our endless Frontier to the same level of development as North America?"[32] Thus one might argue that while it may be true that many have until now desired to be modern, it seems ecologically well-nigh impossible that we will ever get to a stage where every human being will partake equally of the benefits of modernization. There, irrespective of whether or not we have ever been modern, we will perhaps never be modern, or not all of us anyway!

Fair enough. But we are not going to make any headway in climate-policy debates if we fail to understand why the nature/culture division—that Latour, Bennett, Descola, and others rightly consider epistemologically unsound—found a fresh and original articulation in the imagination of the colonized. It is precisely on this question, I think, that postcolonial criticism has some distinctive contributions to make to this discussion.

Unless we understand this dream of the colonized—who had been told to wait for self-rule until they were "modern" enough to deserve it—we will not understand the complaint, made in every colony but voiced famously by Aimé Césaire in the closing paragraph of the first chapter of his book on colonial discourse, that European colonial rule amounted to a promise that was deliberately left unfulfilled: "The proof is that at present it is the indigenous peoples of Africa and Asia who are demanding schools, and colonialist Europe which refuses them; that it is the African who is asking for ports and roads, and colonialist Europe which is niggardly on that score; that it is the colonised man who wants to move forward, and the coloniser who holds things back."[33] All anticolonial nationalisms, as Césaire highlights, were programmatically committed to modernization, the project of making the nation modern. Nehru, Nasser, Mao, Ho Chi Minh, Julius Nyerere, Sukarno, Leopold Senghor, Aimé Césaire—all were radical modernizers pedagogic in their relationship to their respective populations and idealist visionaries of what would turn out to be energy-guzzling human futures. Gigantic in their own national contexts and inspired by a variety of models of economic development ranging from the American to the Soviet ones, these were men who embodied the desires of those in the world who, in the wake of the rise of European nations to world-dominant status, always wanted to be modern.[34] Are they already accounted for, say, in the critical stance that informs Latour's brilliantly polemical and profound work? I think not.

Modernization and the Ethics of the Nature/Culture Distinction

Let me share with you some examples from Nehru's statements to show how spiritual and idealistic was this passionate third-world desire for energy-intensive, mostly fossil-fuel-driven modernization. This was three or four decades before the currents of a consumerist globalization swept through the world and about fifteen years away from the new social movements—including second wave feminism and the environmentalist movements—of the 1970s. Nehru saw from the beginning of his term (1947) as the first prime minister of India that the fundamental problem to address in a country that had seen major famines under British rule till as late as 1943 was the availability of food grains.[35] Irrigation was essential to growing more food, and central to irrigation was the question of power. This made the Himalayan glaciers and all the rivers flowing out of them into India into some kind of a "standing reserve" for Nehru. His first priority, he thought, was to dam the rivers to extract both irrigation water and electricity out of them. At a public meeting in Calcutta in 1949, Nehru spoke of the

> big plans before us: . . . In two or three years we shall successfully complete the river valley projects of Damodar Valley, Mahanadi Scheme, Bhakra Dam and others all over the country, from south to north, and that shall bring lakhs of areas under irrigation. With the completion of canals we shall produce more food and also electricity. So we shall solve our food problem in 5 to 7 years. But we have immediate plans also to solve the food problem. . . . We hope for an extensive and successful agriculture in the Rajasthan desert after it gets canal waters. . . . That shall happen.[36]

The Himalayas where many of the glaciers are receding today have a fascinating presence in Nehru's speeches. They appear at two levels of abstraction—as political and topographical maps in his prime ministerial office, and then as his imagination of them. He liked mountains in a romantic spirit, but the prime minister in him would push all those feelings aside—"I like the Himalayas myself; I like mountains and all that"—to make room for a more extractive vision of the hills: "When I see a map of India and I look at the Himalayan range . . . I think of the vast power concentrated there which is not being used, and which could be used, and which really could transform the whole of India with exceeding rapidity if properly utilized." As a "source of power," the mountains seemed most "amazing," probably "the biggest source . . . in the world—this Himalayan range, with its rivers, minerals, and other resources." That is why all the rivers issuing from the hills had to be "developed" for the progress of the nation. That is why he attached "more importance"—more than what his romantic sentiments urged— "to the development of those big river valley schemes, dams, reservoirs, hydro-electric and thermal power and so forth, which, once released, will simply drive you forward."[37]

This utilitarian but idealist abstraction of the hills would also defeat—at least in the prime minister in him—the scholar who had always displayed a romance of both "world history" and of Indian history in his two major books, *Glimpses of World History*, inspired in part by H. G. Wells, and a text that is still read in classes on the nationalist imagination, his classic *The Discovery of India*.[38] "Look at the map of Asia and of India. It stares at me in my room and in my office, and whenever I look at it, all kinds of pictures come into my mind," he said in a speech to the Central Board of Irrigation in December 1948. What kind of pictures? By his own recounting, the first images that came to his mind were not that of industrial progress but a much gentler picture "of the long past of our history, of the gradual development of man from the earliest stages, of great caravan routes, of the early beginnings of culture, civilization and agriculture, and of the early days when perhaps the first canals and irrigation works were constructed and all that flows from them." But "then," he said, marking an important caesura in his thinking, "I think of the future." A future that, in a manner reminiscent of what Koselleck said about *Neuzeit* or the time of the modern, would derive its horizon of expectation not from the space of historical experience but somewhere else, a *uchronia* (in Derrida's locution).[39] When he thought of the future, Nehru said, his attention would be "concentrated on that huge block of massive mountains called the Himalayas which guard our north-eastern frontier." "Look at them. Think of them," he would exhort his listeners. "I know of no other place in the world which has as much tremendous power locked up in it as the Himalayas and the water that comes to the rivers from them. How are we to utilize it?"[40]

Time and again Nehru would return to this theme. "When I look at the map of India—I look at it very often—it stares me in the face in my office," he said in his opening address at the twenty-third annual meeting celebrating the silver jubilee of the Central Board of Irrigation and Power at New Delhi on November 17, 1952, "I often think not only of the fact that great mountain chain is a boundary of India, . . . not only that it rises up like a sentinel, not only that it has been the inspiration of so much of our culture and thought in the past, but I think also of that mighty chain being a suppressed source of vast energy. The energy flows out in great rivers coming from those mountains and watering the plains of India, running into the sea, then it takes the shape of minerals and the rest of it." And then came his utopian bravura: "So it seems to me, here is a mighty reservoir of energy which if only we could utilize it to full purpose, what could we not do of it?"[41]

* * *

Science and technology would have had to be of central importance to such a vision. Speaking to an Industries Conference in Delhi on December 1947 (four months after independence, that is), Nehru said, "Many things contributed to the winning of the last war, but I think the final reasons were two, the amazing capacity of American industry and scientific research."[42] As he famously did with dams, he designated Indian scientific laboratories, too, as her "modern temples": "I look upon them [scientific laboratories] as temples of science built for the service of our mother land. . . . Service to science is real service to India—no, even to the whole world; science has no frontiers."[43] A year later, on December 5, 1948, addressing the Nineteenth Annual Meeting of the Central Board of Irrigation, New Delhi, he reiterated this faith in science:

> There was a time in the past . . . when it might have been said with some correctness that the world's resources were really not enough to raise the standard of living of the population of the world to the extent desired. Now, I suppose it must be clear to the meanest intelligence that with proper utilization of the present resources of the world—leaving out further development, or even leaving out the world if you like, we can raise the standard of India. This can be shown with a pencil and paper. . . . We have to convert this vast potential into actuality.[44]

We would get the likes of Nehru or Mao or Nasser or Nyerere wrong if we thought of them as pragmatic people expressing a simple and naive faith in technocratic solutions to the problem of energy or water supply. Nehru saw the task of making the nation "advance" as nothing short of a spiritual mission, one that required both idealism and faith on the part of the technocrat—but a faith that went far beyond questions of technological effectiveness. What Nehru's vision called for was faith in both the people of the country and in the project of modernization in the interest of unleashing popular energies in creating a nation. There are some telling anecdotes that Nehru himself recounts. Speaking to the Board of Irrigation and Power in December 1958, he recalled that he went, "four or five years ago," to the Damodar Valley Corporation, where "an enthusiastic young engineer explained to me what they were doing." Nehru was happy to see this man's "interest excited" and noticed that there were "a few hundred men and women [around] carrying baskets of earth on them." He commented,

> I asked the engineer, "Did you explain to them the reasons for what they were doing?" He said, "No." I said, "Then you have not understood your work at all. Your work is to explain to the ordinary worker what he is doing in the scheme." . . . Later I called the hundreds of people who were carrying earth from one place to another. I said, "What are you doing?" They said, "We are taking this basket of earth from here to there." They did not even know the immediate use of their works as part of a big scheme. . . . [Yet] those are the people who are going to profit ultimately when the scheme is ready. It is up to the personnel who are working in the Damodar Valley Corporation to see that the people of the whole area, the village and other places, know what they are doing.[45]

Faith was ultimately about faith in the project of modernization and faith in trusting it to the people of the nation. All the talk about dams and laboratories being "temples" was about creating a secular religion of modernization. "No man can build or construct anything beautiful unless he has faith. See the magnificent cathedrals of Europe ... the embodiment of the faith of the builder," Nehru said to his Irrigation Board in 1948. But "now we live in a different age. . . . [Our] public works should also be fine and beautiful, because there is that faith. So I would like you to work in that faith and you will find that if you work with the faith and that spirit, that will itself be a joy to you."[46] It is not accidental that so many of the speeches I quote here were made to engineers who worked in irrigation and power. "When I read the name of your board, the words 'Irrigation and Power' excite my mind," Nehru remarked in an address to this group in 1952. This is why, he also explained in the same speech, the subject of irrigation or electricity was never "dry or dull" for him—it was "a subject of adventure and excitement and human progress."[47] "I should like you," he further wrote, addressing "not only the big engineer, the middling engineer, but the small engineer," to "convey something of the exciting approach to this problem to the workers there in the field. Make him realise that he is also working with live material even it might be stone or steel and that it will give birth to further life. Let him be the partner in this adventure which you are starting ... [and] other results will follow. . . . The worker and the engineer will also progress and advance and become better men and women."[48]

Of course, this spiritual, ethical, and idealist side of the developmental discourse rings hollow today—at least in an age of jobless growth and intelligent machines, third-world political leaders invoke it in bad faith. The current Indian prime minister, Narendra Modi, authored a book on climate change in 2011 when he was still the chief minister of the state of Gujarat.[49] The rhetoric of the book that has been described as Modi's "green autobiography"—an apt description since every good policy of the state of Gujarat is portrayed in the book as stemming from one person's response to what he saw around him—is strikingly different from Nehru's.[50] Science and technology do not appear here as agents of disruptive, utopian, revolutionary transformation of both spirit and matter. The message throughout the book is of harmony—with two successive chapters carrying headings such as "small is beautiful" and "big is also beautiful."[51] The biggest harmony is, of course, that between ancient Hindu scriptures—the *Vedas*—and modern climate science, whose essentials had all been anticipated in the scriptures. "My views on the complementary relationship between man and nature," writes Modi, "took definite shape when I studied the *Prithvi-Sukta* of the *Atharva Veda* during my college days. The sixty-three Suktas (couplets) composed thousands of years ago, contain a whole spectrum of knowledge which is now being propounded under various scientific, academic and analytical banners during discussions of global warming, damage to earth's environment and the resultant Climate Change."⁵² Indeed, we could not be farther away from Nehru's time and temperament.

But were the leaders of Nehru's generation—all modernizers—merely examples of Naipaul's "mimic men," half shadows of Western or European modernizers, devoid of any originality? Such a judgment would fail to understand the problem of "originality" as anticolonial nationalism poses it—Partha Chatterjee's powerful analysis of this genre of nationalism is instructive here—and would be completely oblivious of Homi Bhabha's deeply insightful reworking of the categories of mimicry and ambivalence in colonial discourse.⁵³ It would be to speak as if postcolonial criticism never happened or had nothing to say to our times.

Latour speaks of "provincializing modernity" as a European task: since Europe brought it about and spread it throughout the world, it is now the European intellectual's task to "provincialize" it, to put it back in its proper place.⁵⁴ But, as I argued in *Provincializing Europe*, Europe was not the only originator of modernity; third-world intellectuals who took heart from what they saw as the universal side of certain European ideas were cooriginators in the process. The global project of modernity got a second and original life in the hands of anticolonial modernizers.

The anticolonial desire to modernize was not simply a repetition of the European modernizer's gesture. In fact, Nehru, like many other nationalists of his generation, often—and self-consciously—addressed this question of mimicking, of simply aping the West. Addressing the Engineering Association of India at New Delhi on December 28, 1962, less than two years before he died, he said, "we have to keep to our roots but at the same time it is equally obvious that no country in the world today can succeed in any sense of the word without understanding what the new world is—the new world of science, technology, etc." This was the dilemma every anticolonial modernizing nationalist faced. Here is Nehru again, continuing on the problem:

> You will see that in the last 200 years or so great differences have arisen in various countries of the world; in the countries of Asia and Europe because Europe had what is called the Industrial Revolution and is continually having that revolution which is changing the life of human beings and the life of groups and societies. And which is not only bringing a measure of well-being to those people . . . [it is also]

strengthening the various nations. . . . We have to find some way of combining the two—a synthesis between what we consider of value in the old and what we consider of value in the new. Mere attempt to copy other countries is not good enough.[55]

This was not the self-image of a mimic man.

India, the third or fourth (depending on how you count) largest emitter of greenhouse gases, is especially vulnerable to the impacts of climate change. Yet what drives politics in India is not the "planet" of planetary global warming but the "globe" of globalization—a revolution of aspiration across classes that has been engendered by political democracy, postcolonial development, and the more recent liberalization of the economy and the media. Up until the time the climate problem became a topic of general discussion, social scientists welcomed this aspirational revolution as a sign of further democratization of the world, a step toward more justice between humans.[56] The history of this outlook must go back to the secular ethic of care for the well-being of fellow citizens that the twentieth-century anticolonial drive toward modernization embodied. Listen once again to Nehru in praise of industrialization from a passage that we quoted above: "[it] . . . is bringing a measure of well-being to . . . people." The very subject of economics, especially welfare economics, emerged in the early part of the twentieth century as this art (or "science," as many economists then believed and still do!) of scaling up and governmentalizing this ethic of care. For instance, introducing the 1929 third edition of his book *The Economics of Welfare*, A. C. Pigou said,

> The complicated analyses which economists endeavour to carry through are not mere gymnastic. They are instruments for the bettering of human life. The misery and squalor that surround us, the injurious luxury of some wealthy families, the terrible uncertainty overshadowing many families of the poor—these are evils too plain to be ignored. By the knowledge that our science seeks it is possible that they may be restrained. Out of the darkness light! To search for it is the task, to find it perhaps the prize, which the "dismal science of Political Economy" offers to those who face its discipline.[57]

Indeed, whether we look at the economist Theodore Schultz's market-based idea of "human capital" that he propounded in February 1959 in his Sydney A. and Julia Teller Lecture at the University of Chicago—which began by acknowledging that "our political and legal institutions have been shaped to keep man free of bondage" and our shared abhorrence of slavery—or at Amartya Sen's later idea of "development as freedom" rooted in giving a person the capability "to promote her ends," we are looking at a family of ideas that go back to European discussions of modernity as freedom that anticolonial leaders like Tagore, Gandhi, Nehru, Fanon, Nyerere, and others renewed and reinvigorated for their own purposes.[58] Economic growth and distribution of welfare seemed to be the best bearer of this ethic of care when such ethic had to be scaled up for communities as large and as impersonal as the nation. We don't understand the Sandhya Chauhans and Kaushilya Devis—or the legitimacy of their voices—today without remembering the desire for modernization and human flourishing that anticolonial nationalisms nurtured and disseminated.

The Difficulty of Being Modern

It is not always possible for humans to transition smoothly from being attached to a human-dominant order of life to being one species among many. While there may be specific areas of life—such as women's reproductive rights—where the language of freedom meshes nicely with what seems ecologically desirable, this cannot be assumed for all aspects of human life, as the story of air conditioning in India demonstrates. The predicament of the political thinker, I suggest, is deeper. The insights of the proponents of the Capitalocene and the posthumanists are important and have to be taken on board, but we need to go beyond the story of original "sins" of capital/labor and nature/culture distinctions to understand the human attachment to "thin descriptions" of nature and thus to modernization. While it could be argued that it is important to inaugurate a regime of politics that took the nonhuman seriously irrespective of whether or not humans could act as spokespersons for the nonhuman, the conversation will not proceed very far without negotiating the desire to be modern that anticolonial ideologies of the twentieth century expressed and that came to shape postcolonial and postimperial formations of politics in so many parts of the world. And these desires were stoked by a global-imperial and expanding universe of travel, exposure, and cosmopolitan conversations that were in turn made possible by the extensive use of energy extracted from fossil fuel. For after all, and for all their criticisms of industrial civilization, where would a Tagore or a Gandhi be if there had not been any railways, steam ships, and printing presses—all manifestations, in their times as in ours, of the enduring power of King Coal and his heirs?

* 5 *
Planetary Aspirations

READING A SUICIDE IN INDIA

On January 17, 2016, Rohith Vemula, doctoral student at the University of Hyderabad, son of a Dalit mother and a low-caste father, took his own life in protest against the university authorities who penalized him for his Dalit student activism. By ending his short and promising life, Vemula made a political-ethical statement with his body; his suicide note reflected on the low-caste/Dalit body itself within a utopian cosmos. The "value of a man" in the society he had lived in—wrote Vemula in his parting note—had always been "reduced to his immediate identity and nearest possibility." "To a vote," he said, or "to a number. To a thing. Never was a man treated as a mind. As a glorious thing made up of stardust. In [e]very field, in studies, in streets, in politics, and in dying and living."[1] Vemula leaves us with two ways of transcending the "untouchable," stigmatized Dalit body: one is by transcending the body altogether, by treating every human being as a "mind" without reference to his or her socially marked body; the other by taking away the "individual" body of the person and connecting it to the material that makes up our universe—ancient atomic and subatomic particles, Vemula's "stardust," that circulate through our and other bodies in the cosmos all the time. The second perspective was not simply a matter of rhetorical flourish. He was a student of science and an avid reader of Carl Sagan; he even quoted Sagan in one of his Facebook posts as saying, "Our species needs, and deserves, a citizenry with minds wide awake and a basic understanding of how the world works."[2] Sagan's reference to the "species" gestures toward a very long-term and collective history of *Homo sapiens* and its journey through time, while his phrase "how the world works" points us toward questions about where humans fit into the story of how the planet functions as a quasi-systemic entity connecting the human with the nonhuman and the living with the nonliving.

It was as if Vemula had read the opening sentences of the last chapter of Kant's *Critique of Practical Reason,* words that also are engraved on the philosopher's tombstone in Kaliningrad.[3] "Two things fill the mind with ever more and increasing wonder and awe," Kant wrote, "the often and the more steadily we reflect on them: the starry heavens above me and the moral law within me." He might not have agreed with Kant's interpretation of the starry heavens—the "former view of a countless multitude of worlds *annihilates* my importance as an *animal creature,*" Kant wrote. Vemula thought of himself as made up of "glorious stardust," not animality. But he might have agreed with Kant's interpretation of the moral law—Vemula's view of man as a "mind"—that "infinitely," wrote Kant, "raises my worth as that of an intelligence by my personality," by revealing "a life independent of all animality[,] ... a final destination ... which is not restricted to the conditions and boundaries of this life but reaches into the infinite." In their thinking, however, they both subordinated to reason the creaturely nature of the body. The two worlds, Kant thought, could be connected by the work of reason.[4] Reading Vemula's dying statements in the light of the Anthropocene hypothesis allows us to reassign importance to the creaturely connections of the human and to demonstrate at the same time the difficulty of bringing these connections within the emancipatory realms of the political.

My point of departure here comes from some stimulating and generative reflections Martha Nussbaum has made on stigmatization and the emotion of disgust as they feature in the philosophy of modern, mainly American, law.[5] It is not the specifics of her arguments that concern me here—though some of her conclusions, such as that we should be skeptical about "relying on [disgust] as a basis for law" since "disgust has been used throughout history to exclude and marginalize groups" may well apply to India—but points where her thoughts touch on the evolutionary psychology of humans. Of course, Nussbaum does not elaborate on these points even when she broaches them, as they are often points she needs to both recognize and bracket in order to proceed with her own exposition. But those are often the points that interest me in this chapter. So it would probably be more accurate to say that my argument forms itself, as it were, on the margins of Nussbaum's text by following up on what she acknowledges but does not feel obliged to pursue.

Nussbaum acknowledges, for instance, that the emotion of "disgust" probably entails elements that belong to a deep history of the human species, including "magical ideas of contamination, and impossible aspirations to purity, immortality, and nonanimality, that are just not in line with human life as we know it." Disgust may have played, she suggests, a "valuable role in our evolution," and it is not just possible but indeed "very likely" that it plays "a useful function in our current daily lives." Perhaps its function of "hiding from us problematic aspects of our humanity is useful: perhaps we cannot easily live with too much vivid awareness of the fact that we are made of sticky and oozy substances that will all too soon decay."[6] "Some self-deception," she writes, "may be essential in getting us through a life in which we are soon bound for death, and in which the most essential matters are beyond our control." Nussbaum leaves it there, as her main purpose in the book is to call for "a society where such self-deceptive fictions do not rule in law and in which—at least in crafting the institutions that shape our common life together—we admit that we are all children [i.e., equals without a father figure] and that in many ways we don't control the world."[7]

Nussbaum also leaves aside—logically, from her perspective—questions of emotions that may be shared between humans and other animals: "I have said that emotions are 'human experiences,' and of course they are that; but most contemporary researchers, and many in the ancient world, also hold that some nonhuman animals have emotions, at least of certain types. . . . I shall leave that issue to one side for now, however, focusing on the human emotions that are the standard material of law."[8] Nussbaum's thoughts are focused on the human alone, and—as with many other liberal thinkers—she thinks of principles that could potentially be applicable to every individual human being irrespective of the total number of humans on the planet. Nussbaum proceeds—rightly, again, from her perspective—from the assumption of "equal worth of persons, and their liberty," for her attention remains focused on human flourishing, that is, on elaborating some "core" legal principles that she considers essential for the flourishing of all individual humans whose lives are governed by institutions that subscribe to liberal principles.[9] Nussbaum's thoughts are anthropocentric by choice.

The "Dalit question" in India, or the persistence in modern Indian institutions of the old problem of "untouchability" in new forms, illustrates at once why both Nussbaum's critique and rejection of disgust as a basis for social management and why Carl Sagan's view of the human body as "stardust" (as Vemula summarized it) are *both* relevant concerns today. They are relevant, but they are also somewhat at odds with each other. In the Brahmanical scheme of things, the body of the "untouchable" person was considered untouchable precisely because it was invested with a certain degree of disgust-arousing significance. This disgust was the emotional source of the marginalization and oppression of the Dalit. From Nussbaum's position, rejecting such a degrading construction of the human body in favor of the individualized body that underwrites the "equal worth of persons" principle is one way to overcome the Dalit's body. And it perhaps speaks to Vemula's complaint that the Dalit could never be seen as someone who had overcome his/her body, and thus, as he put it, demanded to be seen "as a mind." The body as "a glorious thing made up of stardust," however, is a construction that sees the human/Dalit body as connected to everything else in the cosmos, to its ancient past and its present. The view here is neither anthropocentric nor one that individuates the human body. While in Nussbaum's view human flourishing refers to conditions under which all individual humans can potentially flourish, the body as "stardust" dissolves the individual body into some connected view of the physical universe and goes beyond the question of human flourishing. The use of the adjective *glorious* by Vemula in describing this view of the body perhaps signifies the majesty and miraculous nature of the body as it appears at least to Vemula's scientific eyes. He clearly saw this as another powerful way to escape in imagination the limits violently imposed on his "low-caste," Dalit-identified body.

In this chapter, I propose a reading of "the Dalit body"—admittedly an abstract construction about which I will have more to say shortly—by placing such a body at the intersection of the two different traditions of thought that I have collected under the signs of Nussbaum and Sagan. What I have called here the planetary age carries a complexity that marks the present moment in human history. It is this: while we cannot *not* think of human flourishing and questions of justice between humans as we move deeper into the present century, pursuing these questions with no reference to how individual human bodies are connected to nonhuman elements on the planet—both living and nonliving—can in the end imperil human flourishing itself. The overlaps between the literature on climate change and Earth System Science convince me that with the number of humans on the planet today, we need to be increasingly more aware of these connections even as we pursue our own flourishing. That we are made up of "sticky and oozy substances that will all too soon decay" may have to become a part of our everyday awareness. Not only that. The point that not humans but microbial and other small forms of life constitute both by weight and numbers the bulk of life on the planet and are central to the drama of life—from the production of soil to the internal workings of the human body, not to speak of the maintenance of the share of oxygen in the atmosphere—may have to be assigned, as the climate crisis unfolds, the status of a salutary fact that humans will need to keep in mind in thinking about planetary conditions that make our existence and flourishing possible.[10]

The Dalit body, as imagined in the oppressive Brahmanical schema, is marginalized because of its forced contact with death and waste matter; however, it is also one example—bracketing for the moment the relations of oppression that upper castes have built around it—of the human body imagined as intrinsically connected to the nonhuman and the nonliving. We could find similar, and probably a lot more benign, examples in the older religious myths of Native Americans, American Indians, tribes in India and Africa, and of the Australian Aboriginals with the crucial difference, of course, that in the context of caste, Dalits were marginalized and oppressed precisely because of such perceived connections.[11] Rohit Vemula clearly found in the planetary conception of the human body—the human as inextricable from other forms of life and nonlife—an emancipatory horizon of thought. What I do in this chapter is show how difficult it still is to "politicize" this connected figure of the human and why the force of Vemula's emancipatory aspirations remain more poetical than political (in contemporary terms).

The Invisibility of the Dalit Body

The phenomenology of the Dalit body, as Sundar Sarukkai has argued, clearly lies in the Dalit—and the Brahman, too, in a perverse manner— being deprived of something profoundly important to human beings, the touch of other humans.[12] Matters of bodily comportment and performance thus play a crucial role in the history of "untouchability" in South Asia. One cannot theorize "untouchability" without theorizing the body and its cultural location in the history of oppression of Dalits in the subcontinent.

Yet there is a certain kind of forgetfulness about this body that marks the vast and otherwise learned literature on caste and untouchability in India. Symptomatic of this, I now think, was the invisibility of the "Dalit question" even in as self-consciously radical a project as *Subaltern Studies*. Most if not all of Ranajit Guha's examples of acts of physical domination and subordination in everyday life in rural India in his classic book, *Elementary Aspects of Peasant Insurgency in Colonial India* (1983) came from literature on caste, but caste was almost an absent category in his—and later our—analytical framework. It was not as though we did not know about caste and its terrible inequities, but caste was sublimated into the categories "peasant" and "class" in the interest of a historiography that was meant to advance a politics of revolutionary transformation of Indian society, a transformation we understood through the prism of a Marxist outlook, however dissident and democratic its spirit may have been. The subject of humiliation by members of upper castes in everyday Indian life was an embodied subject—sporting a moustache, carrying an umbrella, wearing shoes or breast cloth gave affront to members of dominant groups in particular societies and elicited a violent response of abuse and torture. The humiliated body was marked by caste and its rules of exclusion, yet caste was what we did not discuss in *Subaltern Studies* for a very long time until criticisms from the likes of Kancha Ilaiah made us aware of this serious gap in our intellectual endeavor.[13]

True, traditional Marxist categories are often blind to "caste" and tend to fold it into the category "class," but that problem had already been recognized as such by the time *Subaltern Studies* came to be published. So why did we, academics working on South Asia with most of us having grown up and experienced caste in its multiple manifestations in different parts of the subcontinent, not recognize caste oppression for what it was—a form of oppression whose logic of humiliation and exclusion expressed itself through the materiality of embodied practices? There are, of course, many factors that contributed to this general elision of the centrality of the Dalit body in narratives of Dalit suffering. One could point to the plethora of caste studies in the 1960s and 1970s that aimed at highlighting facts about social mobility within the caste "system" in order to dispute the European canard that the so-called caste system was a straitjacket that held people inevitably confined to the caste (*jati*) into which they were born.[14] The category "caste" belonged here to an emerging discipline of Indian sociology. Dalits and the question of untouchability were folded into the problem of caste, and caste—like race (though many argued caste was not race, and there was ae entire CIBA foundation volume dedicated to this question alone)—was seen as a form of inequality that democracy, socialism, or sheer market or developmental logic were meant to take care of in the end.[15]

There was also an idealistic strain in criticisms of caste-related oppression that portrayed a "spiritual" history of India or Hinduism by emphasizing the emancipatory potential of the Bhakti movement—a devotional form of religion that borrowed antihierarchical elements from both Hindu and Islamic sources—in a gesture calculated to give the egalitarianism of Indian democracy a deep historical genealogy. In modern discussions of this literature, the problem of the body of the Dalit was often converted into a problem of the spirit—a matter of consciously or unconsciously held attitudes that could be spelled out and questioned in religious texts. This was a civilizational narrative of India in which certain Indic texts are seen as having prefigured solutions to problems that the nascent Indian democracy born after 1947 would have to face. One good example are the Patel lectures the famous Sanskrit scholar Raghavan delivered in Delhi in 1964 at the invitation of Indira Gandhi, who was then the minister of information and broadcasting. In these lectures, Raghavan took his audience through an enlightening tour of the various phases of the Bhakti movement from the sixth to the seventeenth centuries in India to paint, as John Stratton Hawley puts it, "a sweeping panorama of India's democratic instincts as they existed before the word 'democracy' was coined."[16]

The blindness to the problem of the body was not just a question of how—that is, through what methods—we discussed caste. There was more to it. I left India when I was 27. In those twenty-seven years of growing up in India, I never heard a single argument—either in school or at home or in social conversations—defending the practice of "untouchability," and yet it remained in everyday life in various forms, some more subtle than others. Of course, knowledge mattered. Knowing about a problem usually leads to action or policy calculated to address it. Hence the various measures India has taken so far to address the problem of untouchability, beginning of course with the remarkable step of pronouncing it illegal in independent India. Yet discrimination—and practices based on age-old assumptions about the body of the Dalit—never really ceases. Why?

Here we need to make a distinction, it seems to me, between, say, particular practices of discrimination and something we may call "prejudice." We become cognitively aware of discriminatory practices and seek to explain them with the various knowledge systems at hand. This is how the various disciplines of history, anthropology, or law would create out of the changing realities of caste their particular object of research and investigation. These knowledge systems, at the same time, also suggest steps for remedial action that may lie in the realms of legislation, economy, politics, or even in consciously held attitudes. Prejudice is something different. It refers to the judgment you make of someone before you consciously judge them—it is in that sense, *pre*-judice, as Gadamer explains in *Truth and Method*.[17] These we imbibe from the earliest phase of our childhood as we come into the symbolic order and as grown-ups explain the world to us and guide us into it, as they necessarily have to. Prejudice becomes part of habitus (to switch from Gadamer to Bourdieu). Oftentimes you see the knowledge/prejudice split in the same person, or, if my logic is right, probably in all of us.

In the interest of time and space, let me illustrate this point with the help of an autobiographical anecdote. I apologize for making autobiography stand in for ethnographic research, but then twenty-seven continuous years in one place is much longer than the time an anthropologist would typically spend in the field over his/her entire life. So perhaps I can claim a certain right to speak as a native-turned-ethnographer. When I was growing up in Calcutta in the 1950s, there was a very famous Bengali poem on the figure of the sweeper included in my school text. It was a stridently antiuntouchability poem beginning with lines that Bengalis of my generation can still recite from memory: "Ke bole tomare, bondhu, asprishya ashuchi?" (Who dares to call you untouchable and impure, my friend?). Satyendranath Datta, the grandson of the famous nineteenth-century rationalist Akshaykumar Datta, wrote it.[18] Datta died young at 39 in 1922. He was an ardent admirer of Gandhi, so the poem probably was composed in the years after Gandhi came back to India permanently in 1915. The poem clearly had a long life. Gandhi began the publication of his journal against untouchability, *Harijan* (1933), with an English translation of this poem by Rabindranath Tagore, and the original Bengali version turned up in my school text some forty years later. My mother, who was a teacher of Bengali literature in a high school, would teach me the poem explaining with much sincerity and fervor the injustice of untouchability and how its every precept did violence to any fundamental principles of human equality and justice. Yet every morning, Lakshman, a Bihari Dalit appointed by the city corporation to sweep our neighborhood clean, would moonlight by cleaning the lavatories of the houses of our streets. (Both of these were standard practices then: the city authorities would invariably appoint Dalits to do sweepers' work, probably a practice even today, and the sweepers in turn would making additional money by taking up private employment during their official working hours.) My parents had a good relationship with Lakshman—he would leave with them his money and other valuables whenever he went home on leave—and never treated him as an untouchable person during these social visits. But every morning when he came into our house as a sweeper wielding a large, wet, and dripping *jhadu* (broomstick) with which he cleaned our lavatory, my mother would scramble to ensure that nothing—no draperies or pieces of furniture—was touched by him or the *jhadu*, producing in the process quite a panicky commotion in the household. Lakshman himself would also walk around assuming a stiff and awkward bodily posture at these moments, taking care to maintain a "proper" distance between his body with the *jhadu* and the furniture and the people of the household so that upper-caste sensitivities about waste matter and pollution were not in any way offended. Richer households would actually build a separate entrance, sometimes even a separate spiraling staircase, for the use of the sweeper.

Growing into my high school years, I came to think of this everyday event as expressive of some kind of hypocrisy on my mother's part. Perhaps she really did not believe in what Datta's poem said, the message of which she would explain to me by way of teaching me the right values of India's egalitarian democracy? I realized later that I was perhaps wrong. My mother was sincere in explaining to me the injustice of untouchability. What was in evidence on Lakshman's entering our house was prejudice in the Gadamerian sense: my mother's deeply Brahmanical sense of her own body was perhaps revolted by the thought that Lakshman and his *jhadu* dripping with water that may been used to clean fecal matter—an extended untouchable body, really—might come into contact with anything in our household. The point was not about hygiene. It was about the body of the Dalit qua Dalit. Formal knowledge of the oppression of Dalits historicizes or sociologizes the figure of the Dalit. Once you know the historical context that aids the exploitation of Dalits, you evolve policies aimed at changing the context of Dalit lives. But prejudice—the judgment you have before you deliberately judge—reproduces a structure with time constituting a very long and stable present.

The Dalit Body as Inscription and Abstraction

The "Dalit body" I mention here is, as I have already said, an abstraction. Since this abstract figure may be mistaken for an essentialist, Orientalist, or static view of the body of the Dalit on my part—as a denial of history, that is—let me begin with a full acknowledgment of the empirical fact that on the ground there is perhaps no one who can correspond to "the Dalit" of my description. On the ground, there are only the bodies of the members so many different *jatis* that were traditionally considered "untouchable." As the Australian scholars Oliver Mendelsohn and Marika Vicziany once observed, "the Untouchables are organized into jatis just as other Hindus are"—"Chamar, Bhangi, Dhobi, Pulaya, Paswan, Madagi are some of the many hundreds of Untouchable jatis scattered through every region of India." And they added, "At the local level everyone knows that there are particular Untouchable castes, rather than Untouchables in general."[19] Dalit intellectuals have themselves related sometimes how much being treated as an "untouchable" was a function of time and place, that is, dependent on the opportunism and selfish interests of the higher castes. A. Shukra (a pseudonym), born in Pune to Panjabi parents belonging to the Ravidasi (worshippers of

Ravidas) caste of Chamars, mentions in an autobiographical essay how the treatment his family received at the hands of their social superiors varied from their time in the village when they would not be allowed to use the water pots of upper castes to the time when he had acquired education and his services were needed by the same social superiors. The "rules of untouchability," he found out, "were complex and hypocritical."[20]

This empirical diversity and the various historical changes are not denied by the conceptual exercise I undertake here. My treatment of the Dalit body is somewhat like Frantz Fanon's treatment of the "black body" in his *Black Skin, White Masks*. The "black man" has no corporeal schema, suggested Fanon, using Hegel and Merleau Ponty, meaning that the "black man" could never forget his blackness; he could not ever forget the color of his limbs or backside, like "humans" do when in everyday being or when they are asleep, say. The black person's sense of his body is always refracted through a third-person consciousness: "In the white world the man of color encounters difficulties in the development of his bodily schema. Consciousness of the body is solely a negating activity. It is a third-person consciousness."[21]

This body of the "black man" that Fanon discussed may have been empirically unavailable for the purpose of verifying his proposition. It is possible that the "black men" Fanon knew, including his own empirical self, were entirely capable of losing all consciousness of the color of their skin while asleep. But that was not Fanon's point. His abstraction, the "black body," was central to a certain structure of racist oppression he wanted to make visible. The "Dalit body" as employed here is a similar construction. I use it to make a point about how we might think about the human body and its completely porous relationship to its so-called environment. The empirical variations in the history of the different groups of Dalits who now constitute India's scheduled castes do not concern me here. For whatever the elements of plurality and variation in the history of untouchability in Indian social history, the body would have to be central to the phenomenon itself. The practices that tend to make a human being "un-touch-able" focus on the body of the person concerned: it is their touch, shadow, their bodily signs and excretions, their food, and so forth, that were and are seen as polluting.

Louis Dumont's classic study of caste, *Homo Hierarchicus*, is helpful here. "It is clear," wrote Dumont, "that the impurity of the Untouchable is conceptually inseparable from the purity of the Brahman.... In particular, untouchability will not truly disappear until the purity of the Brahman is itself radically devalued; this is not always noticed."

Dumont continues to comment on the centrality of the association between the cow and death in the constitution of the defiling nature of the untouchable person:

> It is remarkable that the essential development of the opposition between the pure and the impure in this connection bears on the cow.... The murder of a cow is assimilated to that of a Brahman, and we have seen that its products are powerful purificatory agents. Symmetrically, untouchables have the job of disposing of the dead cattle, of treating and working their skins, and this is unquestionably one of the main features of untouchability.[22]

Dumont's powerful study has been much criticized in the literature on caste, and we do not have to debate either his propositions or his methods. But a sharp memory of the body he describes—mediated sometimes by the reminiscences of a person no less than the great Ambedkar himself—animates Gopal Guru's powerful efforts to conceptualize the experience of being Dalit. "During the Peshwa rule in [early] nineteenth-century Pune," recalls Guru, "the Brahmins forced the untouchables to tie an earthen pot around their neck and a broom around their waist. The pot was to spit in and the broom to erase their footprints that were also considered polluting." Mahars, the untouchable caste that Ambedkar belonged to, were expected to carry sticks with bells attached to them so that the "noise of the bell would communicate the undesirable arrival of untouchables in the main village." This past is not quite dead for Guru. "Thus," he remarks, "the Peshwa rule seems to have developed the prototype of today's biometric techniques," rendering Dalit bodies into inscribed surfaces.[23]

A Reading for the Anthropocene

Let me thus return to the Dalit body that is marked by its involvement with both fecal matter and the skin of dead animals or with death itself (as in the case of the *dom* or the *chandala* of the famous Raja Harishchandra legend that occurs in several *puranas* and influenced Gandhi's thinking). Recall Gyan Prakash's description of "untouchable" bonded laborers in Bihar—the landlords would always ask them to do the first plowing of the land every cultivating season, for the upper castes did not want to risk their bodies by facing the death-dealing matter the earth was meant to give off at the touch of the first plow.[24] The Dalit's body was the buffer between life and death. It absorbed all that could spell death to humans. The prejudice against that body was and is part of the habitus of upper-caste embodied selves.

I do not wish to enter policy or legal debates here first, because I am not competent to do that, and second, because the prejudice against the Dalit body has survived legal and policy initiatives (which is not to devalue these initiatives—we need them). *Subaltern Studies* failed to account for the Dalit because it had no material theory of the body; its "subaltern" was a representative of "insurgent consciousness." But that is not where I want to return. I want to suggest that once you grant me the structure of exclusion—the reaction of disgust it produces in the bodies of "cleaner" castes—we can think of the Dalit body as precisely the body that helps us to think the planet in this age of the environmental crisis that passes by the name of "global warming." To do so, however, we need to get beyond the moves in political philosophy that privilege the abstract, unmarked body either as the carrier of rights or as the ground on which to situate that Marxist category of "abstract labor" so necessary to Marx's critique of capital. Our thoughts on human flourishing perhaps cannot be grounded any more in political thought that focuses on the individual human (as bearers of rights and recipients of welfare) irrespective of the total number of humans on the planet and that brackets all questions of connections between human and other forms of life and their profound relationship to the Earth system processes.

Fanon said—as I have mentioned before—the black person had no "corporeal schema." It is possible for a nonblack person to forget, for instance, what his or her own particular body looks like and be aware in everyday consciousness of just a bodily schema, such as having a vague awareness that he or she had two hands without necessarily remembering or visualizing the color or the shape or the age of the hands. The black person could not do that, for he or she could never forget—even in their sleep—that he or she was black, so deep was the mark that race left on their own embodied sense of themselves. One might be tempted to think likewise of the Dalit body. One could argue that the Dalit person, his or her body always already marked by its proximity to and contact with feces and animals under conditions where the Brahmanical schema of the body dominates, can never experience a general schema of the human body. The Brahman's disgust, as Dumont argues, is inseparable from the stigma the Dalit body bears.

I would, however, resist surrendering completely to such a line of thinking. To put the Brahmin's disgust and the Dalit's closeness to feces and dead animals into an inseparable binary opposition is to remain locked in a kind of humanism that overlooks the live matter in feces and animals, dead or alive—in short, the question of microbes. Since this fact is often forgotten in ontological thinking about the human where the human stands all alone and in abstraction from other life-forms in the world, we could look on the Dalit's body as both an acknowledgment and a reminder, however perverse in its constitution, of all the other living bodies we need to connect with in order to keep our human bodies alive.[25] If we could get out—even in pro-Dalit thought that only focuses on injustice between humans—of anthropocentric thinking, then we could see the Dalit's body as the body that makes us aware of all the networks of connections between different life-forms that enables humans, as a form of creaturely life, to survive. The Dalit's body is itself constructed nonanthropocentrically—it is always human *with* animals, live or dead, and embedded in the world of microbes (with its relationship to the handling of waste). In that sense, the Dalit's is what I might call the planetary body.

In saying this, I do not at all mean to romanticize the vulnerability of the bodies of the poor, be they Dalit or not, who do not have adequate access to health care. Nor do I suggest that we make ourselves vulnerable to diseases and death. There is no "friendly" relationship humans can have to bacteria and viruses that are or become hostile to human life. At the same time, it is true that we owe much of our health to friendly or commensal microbes living in our bodies. My point is about two different but related questions: How do we (re)imagine the human as a form of life connected to other forms of life, and how do we then base our politics on that knowledge? Our political categories are usually imagined not only in profoundly anthropocentric terms but in separation from all these connections. But can we extend them to account for our relationship to nonhuman forms of life or even to the nonliving that we can damage (such as rivers and glaciers)?

Take the human-animal conflict that is ubiquitous in South Asia today. The so-called "monkey menace" in Delhi, caused by habitat loss for monkeys, is a matter of everyday experience. Frequently in India the media carry reports about human-leopard or human-elephant conflicts (as a simple google search will confirm). The question is, contra Hannah Arendt, Can the figure of the refugee remain only human anymore? Should we not think of wild animals such as leopards, monkeys, and elephants that turn up as unwelcome guests in South Asian cities as refugees too? And we have not even begun to think about our relationship to microbial life, though biologists have some definite knowledge of their role in our pasts and futures (viral responsibility for human differences of phenotypes, for instance). However we find a beginning to such thinking, we need to imagine the human not in isolation from other forms of life, in the blinding light of humanism, as it were, but as a form of life connected to other forms of life that are all connected eventually to the geobiology of the planet and are dependent on these connections for their own welfare.

Vemula's emancipatory thoughts—his protest against the oppressions of caste and what in India is called "vote-bank politics"—moved between two perspectives: a liberal-humanist perspective of seeing the human body as unmarked ("never was a man treated as a mind") and a nonanthropocentric perspective derived from science that looks on man as "a glorious thing made up of stardust." The last statement was not a piece of rhetorical flourish nor a figment of romantic imagination but actually a scientific fact on which my Chicago colleague Neil Shubin has written illuminatingly: "Each galaxy, star, *or person* is the temporary owner of particles that have passed through the births and deaths of entities across vast reaches of time and space."[26]

This chasm between the place that astrophysics, geology, biology, and the story of human evolution assign to humans in big histories and that assigned by political thought since the seventeenth century has generally been a matter of pragmatic compartmentalization of knowledge. We know, for example, that humans, apart from being an arithmetic sum of the total number of humans on the planet, are also a biological species, *Homo sapiens*, but that knowledge is usually treated as being of no special political import. But when biodiversity in the world faces, for the first time in its entire history, the bleak prospect of a "great extinction" driven by the activities of one biological species, *Homo sapiens*, the urgency of creating a sense of politics based on this second understanding of ourselves as a species deeply embedded in the history of life dawns on us. But here is the problem that Rohit Vemula thoughts ran up against: we don't know yet how to do that. One might read posthumanists as giving us visions of cosmologies that could help us leap over the chasm between political thought as it exists and political thought as we need it to be. But at present, this is a leap of faith. The chasm exists as the awareness of a deep abyss that acts as the limit to our current human sense of politics. The latter remains focused on individual humans as bearers of rights or as recipients of welfare but never on humans as a totality—one species among many in the larger history of life. This is the chasm that Rohit Vemula pondered in his quest for emancipation but could never cross.

But the failure, if that is what it was, was not Rohit Vemula's alone. Even when political theorists of our time have felt obliged to acknowledge humanity's connections with other forms of life and with the nonliving, they have simply had no intellectual resources within political thought to "politicize" such connections. Consider, for instance, the following passage that occurs early in an otherwise engaging discussion on a possible "political theory of climate change" in Steve Vanderheiden's book on atmospheric justice. It begins with what can be easily recognized as a nonanthropocentric position on the climate crisis, a position that eminently recognizes the fact that humans are embedded in what following Darwin many call "the web of life":

> Carbon is one of the basic building blocks of life on the planet earth, with CO_2 the dominant means by which carbon is transmitted between natural carbon sinks, including living things. In an exchange known as the *carbon cycle*, humans and other animals take in oxygen through respiration and exhale CO_2, while plants absorb and store CO_2, emitting oxygen and *keeping terrestrial life in balance*.[27]

Vanderheiden acknowledges that without the greenhouse gases and "the *natural greenhouse effect*," the planet would be inhospitably cold for life in general and for human life in particular. "While some life," he writes, "might be possible to sustain within a small range of temperature variability beyond that seen since the last Ice Age, the climatic equilibrium produced by 10,000 years of GHG [greenhouse gas] stability is responsible for the development of *all terrestrial life*, and even tiny changes from that equilibrium could throw those ecosystems dramatically out of balance."[28] Vanderheiden is factually wrong since the coming of complex, multicellular life preceded the Holocene by some hundreds of millions of years, but he is right in seeing the modern atmosphere of the planet as an entity shared by different forms of life.

Yet in spite of fully acknowledging that the climate crisis concerns "the balance" of "all terrestrial life" on the planet—whatever such "balance" might mean—and therefore needs be thought of in terms at least of thousands of years, Vanderheiden's questions of justice and inequity circle around problems of human life and human life alone, and problems that are actionable only on much smaller, human measures of time. As he himself says: "While anthropogenic climate change is expected to visit significant and in some cases catastrophic harm on the planet's *nonhuman species*" (emphasis added), his pursuit of issues of climate justice would follow the IPCC in focusing exclusively on "the planet's human habitats and populations." Vanderheiden gives a good, practical reason for this approach: we do not yet know how to compose a global climate regime that would include representation for "animals and future generations"—not to speak of nonanimal life-forms or even the inanimate world. He refers to the work of the political theorist Terrence Ball to argue that even if we represented these groups "by proxies in democratic institutions, giving at least some voice to their interests, . . . they would necessarily remain a legislative minority."²⁹ Thus, it is acknowledged, on the one hand, that "the global atmosphere is a finite good" and is so not just for humans, for it is "vital for the continuation of life on this planet" while being "instrumental for human flourishing" as well. This is the lesson of the sciences. And yet, on the other hand, when it comes to justiciable issues of inequality with regard to climate change, the absorptive capacities of this "one atmosphere"—which, it is acknowledged, "must be shared between *all* the planet's inhabitants"—are divided up *only* among humans ("the world's nations or citizens") with no discussion of what might be the legitimate share of nonhuman forms of life!³⁰ From here it takes only one step to forget nonhuman life altogether and declare global warming to be synonymous with issues of human justice and even to see it as a problem that cannot be remedied *until* issues of human justice are satisfactorily addressed. See how the quotation below moves from a moral recommendation—"concern for equity and responsibility *should not be* dismissed" and so forth—to a conditional statement—"anthropogenic climate change . . . *cannot be genuinely remedied unless*" and so forth—and finally to a statement that posits a relation of identity between global justice and climate change:

> Concern for equity and responsibility should not be dismissed as secondary to the primary goal of avoiding catastrophic climate change, for . . . anthropogenic climate change is also a problem of justice and so cannot be genuinely remedied unless the international response aims to promote justice [including the poor nations' "right to develop"]. . . . Global justice and climate change [are] . . . manifestations of the same set of problems.³¹

My second example comes from a reputed political thinker of our time—the theorist of republicanism Philip Pettit. In his acclaimed book on republicanism, Pettit adduces some "decidedly anthropocentric" reasons for "why we should be concerned about other species and about our ecosystem generally." But notice how humanity—a "we"—occurs in his prose as two distinct and unconnected figures and even this lack of connection goes unremarked. "The ecosystem, with the other species of animals that it contains, offers us our place in nature; it is the space, ultimately, where we belong," writes Pettit. But this "we" is an arithmetical sum of a collection of individuals, a sigma function, as it were, drawn over the basic activities that define the individual human: "We are what we eat. And equally we are what we breathe, we are what we smell, we are what we see and hear and touch." Clearly, eating, breathing, smelling, seeing, hearing, and touching are all activities that could be carried out only by the individual human body. But the same Pettit also writes, "We live in physical, biological, and psychological continuity with other human beings, with other animal species, and with the larger physical system that comes to consciousness in us."[32] "Physical, biological, and psychological *continuity*": this second "we" then is not an arithmetic sum of individual humans. It is a figure of "continuity" that connects us to other species and to processes we may consider planetary. It "comes to consciousness in us" and yet we cannot dispense with the figure of the individual and the autonomous human subject who remains the mainstay of political thought. This problem is peculiar to political thought, as our political institutions are in the end profoundly anthropocentric. Anthropologists, on the other hand, have struggled thoughtfully to bring to life in their prose some of the critical functions shared between humans and nonhumans.[33]

Pettit's thoughts therefore also lead us to the same chasm that Vemula pondered. We now know that the story of human flourishing—the uneven narrative of modernization that has in its sight every individual human—has now run up against a deeper story about humans, our collective unconscious history as biological species that, in the history of life on this planet, is the first to have successfully dominated its entire landmass and, indirectly, even large parts of the oceans. How do we bring both versions of the human—in Vemula's terms, "every human being treated as a mind" and the same person as "star dust"—together to constitute a new kind of political thought? Until we can answer this question satisfactorily, being modern will remain a difficult position to occupy in times that are simultaneously both global and planetary in the senses in which these words have been used in this book.

This is why Vemula's cosmological imagination of emancipation remains, in the end, poetical—because the thinkers of the political do not yet know how to construct the political on the basis of the understanding of the human body that several branches of science give us: that it is porous in its boundaries and remains a zone through which other forms of the living and the nonliving necessarily traffic.[34] As a reader of Bengali literature, I find the poignant poetry of Vemula's thoughts reminiscent of a letter that the poet Rabindranath Tagore—with a very different background to Vemula's—once wrote to Ramedrasundar Tribedi (1864–1919), a pioneer popularizer of science in Bengal. The letter was written about a year before he, Tagore, was awarded the Nobel Prize for literature. It is dated February 29, 1912.[35] From evidence internal to the letter, the following would appear to be its background. Tribedi was preparing some of Tagore's old letters for publication as a book (*Chinnapatrabali*) and appears to have edited out a sentence from one of them.

"There was a day," Tagore had written, "when I grew into a leafy tree on a young and moist earth bathed in seawater."[36] This was a sentence that Tribedi found unworthy of inclusion because its publication before in an excerpted form in a Bengali magazine had caused unending mirth in a journal hostile to Tagore, *Sahitya*. Tribedi was trying to protect Tagore from future ridicule. Tagore's letter to Tribedi was an attempt to exercise—with his characteristically gentle sense of humor—the author's right to protest while admitting the editor's prerogatives in the "execution of [his] duty" (this phrase occurs in English in the letter).

Tagore's protest—much like Vemula's but decades before Carl Sagan and his cosmology were around—ran as follows.

> You have raised the editorial axe against my memories of [having been once a] tree. But this [action of yours] is not like the pruning of unnecessary branches, it is striking at [the root of my] life. Because this is my inmost realization. Within my life there is a secret memory of the life of trees. I can acknowledge it only because I am a human being today. Why only trees? Within me are deposited memories of the entire material world. All the vibrations of the universe bring thrills of kinship to my entire body—the silent and ancient exuberance of trees and creepers have found today a language in my life—why else would I feel called upon to celebrate the Spring right now when budding mangoes on trees seem to be intoxicated with a joyous spirit? Why would you not let me express the tremendous sense of joy [coming from] water, land, trees, and birds that [keeps] coursing through me? Why? Lest people should make fun of me?

He then added, "Whenever, at auspicious moments, the realization that I am here together with the sun, the moon, the stars, and the land, rocks, and water rings out in my mind with the clarity of a musical note, my body and mind experience the intimate thrills of a vast existence. This is not me poeticizing, this is my nature [speaking]. It is out of this nature that I have written poems, songs, and stories. I do not feel the slightest bit of shame about this. It is because I am a human being that the entire truth of the nonliving and the living finds itself in a state of completion in my existence."

Much of this may seem a Heideggerian-sounding proposition about the specialness of man—only in the human does the world find its own consciousness. But Tagore complicates the thought by striking a different note in the end. He remained a stranger to the elements with which he was one: "the waves in my bloodstream dance to the rhythm of the waves in the sea—but the waves of the sea cannot recognize me. . . . The joys of my life blend in with those of the trees but the trees do not know me. They do not carry my memory [as I do of them]. But what is there to laugh at in all this?"[37]

Tagore's and Vemula's visions were not the same. Vemula's idea of himself as a "glorious" piece of stardust did not assign the "glory" to humans. The glory belonged to the cosmos. Tagore's was a celebration of his existence as a human in the cosmos. Theirs were both expanded visions of the human, visions that connected humans both to the living and the nonliving. In Tagore's time, this was a poet's vision; in Vemula's emancipatory reading, Carl Sagan's astrophysics bought him glimpses of a figure of the human liberated from the indignities suffered by the Dalit. Both were thinking at the limits of political thought while responding in their human souls to the invitations of the planetary.

* 6 *
In the Ruins of an Enduring Fable

The year 2015 was the first when the average surface temperature of the world rose by 1°C above the preindustrial average, thus taking us closer to the threshold of a 2° rise, a Rubicon we are told we must not cross if we are to avoid what United Nations Framework Convention on Climate Change (UNFCC) of 1992 described as "dangerous anthropogenic interference with the climate system."[1] The year 2016, as one meteorologist put it, was "off the charts" as far as global warming was concerned.[2] The historian Julia Adeney Thomas remarked in 2014 that the idea of being "endangered" could not be a purely scientific idea, for the planet has been through many other episodes of climate change — and five great extinctions of species — before.[3] *Dangerous* here is indeed a word that scientists, politicians, and policy makers use as concerned citizens of the world, glossing *danger* as a threat to human institutions. In Thomas's words,

> historians coming to grips with the Anthropocene cannot rely on our scientific colleagues to define "the endangered human" for us. . . . It is impossible to treat "endangerment" as a simple scientific fact. Instead, endangerment is a question of both scale and value. Only the humanities and social sciences, transformed though they will be through their engagement with science, can fully articulate what we may lose.[4]

Indeed, one of the first general books to be written on the problem of anthropogenic climate change around the time of the publication of the Fourth Assessment Report of the Intergovernmental Panel on Climate Change (IPCC), Tim Flannery's *The Weather Makers*, pointed out that the entity to which climate change posed a real threat was human civilization as we have come to understand and celebrate it.[5] *Civilization*, of course, is a value-laden and therefore contested word that humanities scholars in recent decades have done much to demystify.[6] I bring up the point here simply to show how central the concerns of the humanities and the human sciences have been to defining one of the gravest problems humans face in the twenty-first century. The point is underlined when the moral philosophers such as Peter Singer describe climate change as the "greatest ethical challenge" ever faced by humanity.[7] True, we could not define "human-induced planetary climate change" except with the help of big science; and, true, the problem of the "two cultures" of the sciences and the humanities remains.[8] But the questions of justice that follow from climate-change science require us to possess an ability that only the humanities can foster: the ability to see something from another person's point of view. The ability, in other words, "to imagine sympathetically the predicament of another person."[9]

This moral demand on humans today acquires an additional twist from the thought that, seen in a long-term perspective, unabated global warming may very well accelerate the already growing rates of human-induced extinction of nonhuman species, with unhappy consequences for humans themselves. Voices have been raised, including that of Pope Francis, recommending that human justice be extended not just to animals that crossed a certain threshold of sentience (as animal liberationists once argued) but to the entire world of natural reproductive life—what Aristotle called the *zoe*. This proposition that in effect subjects the domain of biological life to the work of the moral life of humans marks, I argue, a critical turning point for the humanities today, as it departs radically from a tradition—inaugurated by, among others, Immanuel Kant—that made a strict separation between our "moral" and "animal" (i.e., biological) lives, assuming that the latter would always be taken care of by the natural order of things. This separation, after all, is what has buttressed for more than a century the much-critiqued gap between the humanities and the physical or biological sciences. Strands of environmentalist thought have questioned and on occasion attempted to close this chasm, but the gap persists and has not been easy to overcome.

To ask, as we do today, how humans might use the resources of their moral capacity to regulate their life as a biosocial species among other species is to bring within the ambit of human moral life something that has always lain outside of its scope: the history of natural life on the planet. This problem is not adequately answered by what has been written on extending the conception of rights to certain animals because, first, the number of animals considered in this discussion is limited by a "sentience threshold," and second, because—as we know today—the bulk of life on the planet is microbial.[10] The assumption—made since at least the Enlightenment and still prevalent in many social science disciplines, including branches of mainstream economics and political thought—that the planet's biosphere will take care of our "animal life" while we struggle in search of a collective moral life without regard to our collective life as a biological species is now under severe strain. This has serious implications for the humanities, which have traditionally served as the domain for the discussion of moral issues in separation from biological life. I argue this by looking first at some relevant writings of Kant in the context of discussions on climate change and possible human stewardship of life on the planet and then engage, in conclusion, with the work of Bruno Latour to show where his thoughts indicate a way forward.

Two Narratives of Climate Change

Let me begin with the two dominant approaches to the problem of climate change.[11] One approach is to look on the phenomenon simply as a one-dimensional challenge: How do humans achieve a reduction in their emissions of greenhouse gases (GHGs) in the coming few decades? Climate change is seen in this approach as a question of how best to source the energy needed for the human pursuit of some universally accepted ends of economic development so that billions of humans are pulled out of poverty. The main solution proposed here is for humanity to make a transition to renewable energy as quickly as technology and market signals permit. The accompanying issues of justice concern relations between poor and rich nations and between present and future generations: What would be a fair distribution of the "right to emit GHGs"—since GHGs are seen as scarce resources—between nations in the process of this transition to renewables? The question of how much sacrifice the living should make as they curb emissions to ensure that unborn humans inherit a better quality of life than that of the present generation remains a more intractable one, its political force reduced by the fact that the unborn are not present to press their case. "The nonexistent has no lobby," as Hans Jonas once remarked, "and the unborn are powerless."[12]

Within this broad description of the first approach, however, are nested many disagreements, ranging from capitalist to noncapitalist utopia of sustainable futures. Most imagine the problem to be mainly one of replacing fossil-fuel-based energy sources by renewables. Some others—on the left—would agree that a turn to renewables is in order but would still argue that because it is capitalism's constant urge to "accumulate" that has precipitated the climate crisis, the crisis itself provides yet another opportunity to renew and reinvigorate Marx's critique of capital. And then there are those who think of actually scaling back the economy, degrowing it, and thus reducing the ecological footprint of humans while designing a world characterized by equality and social justice for all. Still others think — in a scenario called "the convergence scenario" — of reaching a state of economic equilibrium globally whereby all humans live at more or less the same standard of living. The role of the humanities is confined here mainly to climate-justice issues with both political economists and philosophers (both in the Rawlsian and utilitarian traditions) contributing to relevant discussions.[13] For all its shortcomings, however, the reduction of the climate crisis to the problem of renewable energy has the advantage that we can develop frameworks of both policy and politics around it.

One can also, however, see climate change not simply on its own but as part of a family of interlocking problems. Exponential population increase, food insecurity, water scarcity, expansion of resource industries and an increase of economic inequalities contributing to human-animal conflicts, habitat loss for other species, GHGs emissions, and so on — all of these are planetary in scope and speak to the fact of an overall ecological overshoot on the part of humanity that affects the distribution of natural life on the planet. Global warming then seems more like a shared predicament for all humans — not to speak of other species — than a problem that is simply a question of switching to renewables. Then there is the knotty question of human "agency" that many scientists have underlined, the new geophysical agency of humans on a scale that has allowed them already to change the climate of the planet for the next one hundred thousand years, putting the next ice age off by anything between fifty and five hundred thousand years.[14] Within this perspective that looks both into deep pasts and deep futures, a very particular challenge opens up for the imagination of modernity. After all, if the problem of planetary climate change arises out of our need to consume more energy than before, then the excess GHGs in the atmosphere could easily be looked on as the resultant "waste" that cannot yet be properly recycled in the time frame suitable for human flourishing (the planet being much too slow for human needs!). Since this human "waste" affects other life-forms — by acidifying the oceans or raising the average surface temperature of the planet — the crisis requires us to do something that the humanities train us to do: "imagine sympathetically the predicament" of others, the relevant "others" here including not just humans but nonhumans as well.

It is, of course, not the physical phenomenon of warming alone that caused this shift in our moral orientation. If one could imagine someone watching the development of life on this planet on an evolutionary scale, they would have a story to tell about *Homo sapiens* rising to the top of the food chain within a very, very short period in that history. "If we imagine," writes John Brook in his masterful study of human history and climate change, "the 5 million years of human evolutionary times as a twenty-four hour period, the entire 300,000 years of modern humanity comprises about an hour and a half, the 135,000 years since modern humans have left Africa comprise about a half hour, and the 12,000 years since the end of the Pliestocene . . . slightly more than four minutes. . . . Not until about 6,000 years ago, more than half the total time elapsed since the end of the Pliestocene, was much of humanity on a clear course towards agriculture." In addition, he remarks, "Viewed from the long history of the earth system . . . , the rise of settled agriculture seems simply a single phase in the brief, explosive eruption running from the emergence of modern humans and their global colonizations and intensifications to our present high-technology, overpopulated, climatically unbalanced condition."[15]

The more involved story of rich-poor differences would be a matter of finer resolution in the big history that Brooke recounts. As I have said elsewhere, the ecological overshoot of humanity requires us to both zoom in to the details of intrahuman injustice—otherwise we do not see the suffering of many humans—and to zoom out of that history—or else we do not see the suffering of other species and, in a manner of speaking, the suffering of the planet.[16] Zooming in and zooming out are about shuttling between different scales, perspectives, and different levels of abstraction. One level of abstraction does not cancel out the other or render it invalid. Nor does this separation of levels deny the point that in our everyday life we sometimes enjoy the geological agency of humans without knowing or calling it by that name (see the introduction). But my point throughout this book has been that the human story can no longer be told from the perspective of the five hundred years (at most) of capitalism alone.

Humans remain a species in spite of all our differentiation. Suppose all the radical arguments about the rich always having lifeboats and therefore being able to buy their way out of all calamities including a great extinction event are true. And imagine a world in which some very large-scale species extinction has happened and that the survivors among humans are only those who happened to be privileged and belonged to the richer classes. Would not their survival *also* constitute a survival of the species (even if the survivors eventually differentiated themselves into, as seems to be the human wont, dominant and subordinate groups)?

The ecological overshoot of humanity does not make sense without reference to the lives of other species. And in that story, humans are a species too, albeit a dominant one. This does not cancel out the story of capitalist oppression. Nor does it amount to the claim that any one particular discipline now has the best grip on the experience of being human. Biology or some other science that misses out on the existential dimension of being human will never capture the human experience of falling in love or feeling love for God in the same way that poetry or religion might. A big brain gives us a capacity for cognition of that which is really big in scale. But it also gives us our deeply subjective experience of ourselves and our capacity to experience our individual lives as meaningful. We cannot produce a consilience of knowledge. But surely we can look on ourselves and on the human story from many perspectives at once.

The phenomenon of the rise of humans to a position of dominance—due, perhaps, to the development of a big brain that has helped humans over tens of thousands of years to create attachments and affiliations to imagined communities far beyond the face-to-face scale of kin group or band—is now seen by many to have taken place over a very long historical period reaching back to times that Daniel Smail describes as our "deep history."[17] The Israeli historian Yuval Noah Harari explains the issue well in his book *Sapiens: A Brief History of Humankind*. "One of the most common uses of early stone tools," writes Harari, "was to crack open bones in order to get to the marrow. Some researchers believe that this was our original niche." Why? Because, Harari explains, "genus *Homo*'s position in the food chain was, until quite recently, solidly in the middle."[18] Humans could eat dead animals only after lions, hyenas, and foxes had had their shares and cleaned the bones of all the flesh sticking to them. It is only "in the last 100,000 years—with the rise of *Homo sapiens*," says Harari, "that man jumped to the top of the food chain."[19] This has not been an evolutionary change. As Harari explains,

> Other animals at the top of the pyramid, such as lions and sharks, evolved into that position very gradually, over millions of years. This enables the ecosystem to develop checks and balances.... As the lions became deadlier, so gazelles evolved to run faster, hyenas to cooperate better, and rhinoceroses to be more bad-tempered. In contrast, humankind ascended to the top so quickly that the ecosystem was not given time to adjust.[20]

Harari mentions an additional significant fact. As a result of their quick ascent to the status of top carnivore, humans themselves, writes Harari, "failed to adjust." He adds, "Most top predators of the planet are majestic creatures. Millions of years of domination have filled them with self-confidence. Sapiens by contrast is more like a banana republic dictator."[21]

The human ecological footprint, we can say, further increased with the invention of agriculture (more than ten thousand years ago but intensifying in the next few millennia) and then again after the oceans found their present level (about six thousand years ago) and we developed our ancient cities, empires, and urban orders while moving to every part of the planet. It increased yet again over the last five hundred years with European expansion and colonization of faraway lands inhabited by other peoples and the subsequent rise of industrial civilization. But it dramatically expanded after the end of the Second World War when human numbers and consumption rose exponentially thanks to the widespread use of fossil fuels not only in the transport sector but also in agriculture and medicine allowing, eventually, even the poor of the world to live longer—though not healthy—lives.[22]

Scholars have carried forward the notion of "overshoot"—"instances in which populations of organisms so changed their own environments that they undermined their own lives"—that William R. Catton Jr. put forward in a book of that name in 1980.[23] The literature on animal liberation/rights that extends the human moral community to include (some) animals recognizes issues of both cruelty to animals and the overshooting of human demands for consumption.[24] Scholars working on human-induced species extinction in the context of anthropogenic climate change have long recognized the "overreach" that humans have achieved, often to their own detriment, in the various ecosystems they inhabit.[25] In addition, well-known arguments about "the great acceleration" and "planetary boundaries" that some earth scientists and other scholars have put forward are statements, precisely, about ecological overshoot on the part of humans. As one of the authors of the "great acceleration" thesis put it, "the term 'Great Acceleration' aims to capture the holistic, comprehensive and interlinked nature of the post-1950 changes simultaneously sweeping across the socioeconomic and biophysical spheres of the Earth System, encompassing far more than climate change."[26] Their data document exponential rise in human population, real GDP, urban population, primary energy use, fertilizer consumption, paper production, water use, transportation, and so on—all happening after the 1950s. And there is corresponding exponential rise in "Earth system trends" to do with the emission of carbon dioxide, methane, nitrous oxide; ocean acidification; loss of stratospheric ozone, marine fish culture, shrimp aquaculture, tropical forests; terrestrial biosphere degradation, and so forth.[27] Similarly, the idea of nine "planetary boundaries" that humans should avoid crossing that was put forward in 2009 by Johan Rockström and his colleagues at the Stockholm Resilience Center was also an exercise in measuring human ecological overreach.[28] Some Earth system scientists reported recently that "the present anthropogenic carbon release rate [around 10 petagrams C per year; 1 petagram = 10^{15} grams] is unprecedented during the [entire] Cenozoic (past 66 Myr)" and that "the present/future rate of climate change and ocean acidification is too fast for many species to adapt" and will likely result in "widespread future extinctions in marine and terrestrial environments." We are, effectively, in "an era of no-analogue state, which represents a fundamental challenge to constraining future climate projections."[29]

Not only have marine creatures and many other terrestrial species not had the evolutionary time needed to adjust to our increasing capacity to hunt or squeeze them out of existence, our GHG emissions now threaten the biodiversity of the great seas and thus endanger the very same food web that feeds us. Jan Zalasiewicz and his colleagues on the subcommittee of the International Stratigraphy Commission charged with documenting the Anthropocene point out that it is the human footprint left in the rocks of this planet as fossils and other forms of evidence—such as terraforming of the ocean bed—that will constitute the long-term record of the Anthropocene, perhaps more so than the excess carbon dioxide in the atmosphere. If human-driven extinction of other species results—say, in the next few centuries—in a great extinction event, then even the epoch-level name of the Anthropocene may be too low in the hierarchy of geological periods.[30] The music historian and theorist Gary Tomlinson, writing recently in the context of climate change, has summed up the problem nicely from an Earth system point of view:

> Across millions of years of biocultural evolution . . . , certain systems remained *outside* the feedback cycles of hominin niche construction. Astronomical dynamics, tectonic shifts, volcanism, climate cycles, and other such forces were in essence untouched by human culture and behavior (or if touched, touched in a vanishingly small degree). In the language of systems theory, all these forces were in effect *feed-forward* elements: external controls that "set" the feedback cycles from without, affecting the elements within them but remaining unaffected by the feedback themselves. . . . The Anthropocene . . . registers a systemic rearrangement in which *systems that had always acted as feed-forward elements from outside human niche construction have been converted into feedback elements within it*.[31]

Viewed thus, as Zalasiewicz says in the concluding paragraph of a recent essay, "The Anthropocene—whether formal or informal—clearly has value in giving us a perspective, against the largest canvas, of the scale and the nature of the human enterprise, and of how it intersects ('intertwines' now, may be a better word) with the other processes of the Earth system."[32] Anthropogenic climate change is therefore not a problem to be studied in isolation from the general complex of ecological problems that humans now face on various scales—from the local to the planetary—creating new conflicts and exacerbating old ones between and inside nations. There is no single silver bullet that solves all the problems at once; nothing that works like the mantra of transition to renewables to avoid an average rise of 2°C in the surface temperature of the planet. What we face does indeed look like a wicked problem, a predicament. We may be able to diagnose it but not "solve" it once and for all.[33]

Modernity and Kant's Geology of Morals

If, as I have claimed, the challenge posed to our moral life by the scale of problems created by our animal life (i.e., humans as consumers, as *animal laborans* in Hannah Arendt's phrase) makes a breach in the assumed separation of our "moral" and "animal" lives and demands of us that we find "moral" solutions to problems created by "natural history" of the human species, then clearly the human sciences, and in particular the humanities, face a novel task today. For it was this very separation between the animal and moral life of the human species that underlay, for a large part of the twentieth century, the separation of the human from the physical and biological sciences.[34] The subject deserves more research. But older readers will remember how vociferously—and oftentimes acrimoniously—sentiments in favor of this separation were voiced when in 1975 Edward O. Wilson published his book *Sociobiology*, making some strong claims about connections between biology and culture and managing to infuriate in the process Marxists and social scientists of many other persuasions.[35]

The enduring importance of the assumed separation of the moral life of humans from their animal or creaturely life in post-Enlightenment narratives of modernity is perhaps best studied with reference to a fable that Immanuel Kant spelled out in a minor essay called "Speculative Beginning of Human History" published in 1786. The opposition between the animal life of the human species and its moral life was at the heart of this essay. The essay provides a fascinating reading of the Biblical story of Genesis and the question of man's dominion over the earth.[36] The aim of Kant's exercise was to bring "into agreement with one another and with reason" what he saw as "the oft misunderstood and seemingly contradictory claims of the esteemed J.-J. Rousseau":

> In his works, *On the Influence of the Sciences* and *On the Inequality among Men*, he [Rousseau] displays with complete accuracy the inevitable conflict between culture and the human race as a *physical* species whose every individual member ought fully to fulfill its vocation. But in his *Emile*, in his *Social Contract*, and in other works he seeks to answer this more difficult question: how must culture progress so as to develop the capacities belonging to mankind's vocation as a *moral* species and thus end the conflict within himself as [a member of both a] moral species and a natural species.[37]

Kant regarded this conflict itself—engendered within man by the human species possessing at the same time both a "physical/natural/animal" (these words are used in the same sense in his essay) life and a moral life—as a decisive influence on human history. For "impulses to vice" arose from "natural capacities" that were given to man "in his natural state"; they necessarily conflicted with "culture as it proceed[ed]." "The final goal of the human species' moral vocation" could not be reached until "art so perfected itself" that it became, in Kant's words, a "second nature."[38]

Many in the vast literature on Kant have discussed the philosopher's answer to the Rousseau puzzle, some tracing certain critical elements in his answer back to ancient principles including those postulated by Aquinas.[39] My purpose here, however, is not a historical excavation of the roots of Kant's thoughts but to reconstruct Kant's argument in order to explicate how precisely he sought to understand the relationship between the animal and the moral aspects of the human being. Kant began his essay by explaining why he could take the liberty of reading the story of Genesis speculatively while clarifying that the speculative was not the same as the "fictional."[40] Speculation could be "based on experience," but the experience in question was that of "nature," something that, for Kant, remained constant in its essential structure. So if human history were a history of freedom, then a statement about its "first beginnings" could be read speculatively (i.e., guided by reason) if we based ourselves on our experience of nature (constant by definition) and only in so far as the beginnings in question were made by nothing other than nature itself. As Kant put it, "A history of freedom's first development, from its original capacities in the nature of man, is therefore *something different* from the history of freedom's progression, which can only be based on reports," and thus become the historian's province.[41]

Kant, of course, made certain assumptions about this original condition of humans so that "one's speculation [would] not . . . wander aimlessly." He took a certain figure of the human for granted—"one must make one's beginning something that human reason is utterly incapable of deriving from any previous natural causes"—and hence began "not with [human] nature in its completely raw state" but with "man as *fully formed adult* (for he must do without maternal care)." He also assumed "man" to actually be "a *pair*, so that [man] can propagate his kind," and the pair had to be "*only a single pair*, so that war does not arise, as it would if men lived close to one another and were yet strangers." This latter assumption, it seemed to Kant, ensured that "nature might not be accused of having erred regarding the most appropriate organization for bringing about" what Kant saw as "the supreme end of man's vocation, sociability" (for the desire to socialize would be maximized by "by the unity of the family from which all men should descend"). Besides, he made some further assumptions to keep his speculative logic straight: "the first man could thus *stand* and *walk*; he could *talk* (Gen. 2:20), even *converse*, i.e. speak in coherent concepts (v.23), [and] consequently, *think*." This threshold of assumptions regarding human skills, he reasoned, would allow him "to consider only the development of morality in [man's] actions and passions." Having thus reconstructed this original pair of humans, Kant placed them squarely in the middle of what we might today see as the geological Holocene period with considerable advances already made in "human civilization": "I put this pair in a place secured against attack by predators, one richly supplied by nature with all the sources of nourishment, thus, as it were, *in a garden*, and in a climate that is always mild."[42] Kant did not know this, but the "man" of his assumptions could have existed only after the last ice age was over!

Kant's "man" began his journey completely absorbed in the animal life of the species when instinct alone—"that *voice of God* that all animals obey"—"first guided the beginner." But by the time Kant has the human being in his sights, reason, a faculty somewhat beyond animal life and yet put in place by some design of nature, had already begun to "stir" and "cook up" in humans—in partnership with a companion human faculty, imagination—"desires for things for which there is . . . no natural urge," with the result that "man became conscious of reason as an ability to go beyond those limits that bind all animals."[43] A critically important discovery followed: "[man] discovered in himself an ability to choose his own way of life and thus not be bound like other animals to only a single one."[44] The deepening of this "inner" propensity gave man the capacity to refuse desires that were merely animal—thus developing the ability to love. "*Refusal*," wrote Kant, "was the feat whereby man passed over from mere sensual to idealistic attractions, from mere animal desires eventually to love and, with the latter, from the feeling for the merely pleasant to the taste for beauty." This, together with the development of a sense of "decency," "gave the first hint of man's formation into a moral creature," a small beginning that for Kant was "nonetheless epochal."⁴⁵ Reason also led humans to "the reflexive *expectation of the future*" and then to a height that raised "mankind altogether beyond any community with animals" enabling humans to conceive of themselves—"though only darkly"—as "the true *end of nature*." Humans could now see that the pelt of the sheep "was given by nature" not for the sheep but for them. Their dominion over the earth that Genesis speaks of had thus begun. But this also led to the idea of equality of all humans–"[men] must regard all men as equal recipients of nature's gifts"—and, more importantly, to the idea that "man became the *equal of all* [other] *rational beings*, no matter what their rank might be (Gen. 3:22), especially in regard to his claim *to be his own end*."⁴⁶ This formulation is, of course, a close cognate of the famous Kantian dictum regarding treating every human being not instrumentally but as an end in himself or herself.⁴⁷

Kant was acutely aware that this "portrayal of mankind's early history" revealed "that its exit from . . . paradise . . . was nothing but the transition from the raw state of merely animal creature to humanity, from the harness of the instincts to the guidance of reason—in a word, from the guardianship of nature to the state of freedom."⁴⁸ This, as Kant explains, had to be the story of a fall, morally speaking. Before reason stirred in the human breast, "there was neither a command nor a prohibition and thus no transgression either." But reason could ally itself "with animality and all its power" and thus give rise to "vices of a cultivated reason" (to produce wars, for instance). "Thus, from the moral side," writes Kant, "the first step from this last state [the state of innocence] was a *fall*; from the physical side, a multitude of never-known evils of life [natural disasters, hardship], thus punishment, was the consequence of the fall."⁴⁹ Much of human history as we know it followed from the fall: there was hardship, inequality—"that source of so many evils, but also of everything good"—wars, and humans getting "drawn into the glistering misery of the cities."⁵⁰ But this also complicated the role of reason in the story of human freedom. Humans could use reason in a way that hastened the vocation of their species—a species designated, according the Genesis story of "man's" dominion, "to rule over the earth, and not as one designated to live in bovine contentment and slavish certitude."⁵¹ But reason did not straightforwardly guide humans toward recognition of their vocation (though Kant in other essays will explain why humans would nevertheless end up fulfilling their destiny).

Kant would thus write, "The history of *nature*, therefore, begins with good, for it is God's work; the history of *freedom* begins with badness, for it is *man's* work."[52]

* * *

The key to human beings' success was "to be content with providence," wrote Kant in concluding this essay.[53] But this was precisely what was never easy for humans to do. Providence worked through what humans considered adversity: wars (that in the end generated "respect for humanity from the leaders of nations"), brevity of life (that guaranteed that improvement accrued to the species and not to individuals), and the absence of a golden age of all leisure and no toil.[54] As Kant put it: "Contentment with providence and with the course of human things as a whole, which do not progress from good to bad, but gradually develop from worse to better; and in this progress nature herself has given everyone a part to play that is both his own and well within his powers."[55]

The late Kant would anticipate, repeat, elaborate on, and develop these basic points in the third *Critique* (the section on teleological judgment) and in several essays including "Idea for a Universal History with a Cosmopolitan Intent" (1784) and "On the Proverb: It May Be True in Theory, But Is of No Practical Use" (1793). Here is Kant, in the third *Critique*, for example, on the subject of the separation of the moral life of humans from their natural history.

> External nature is far from having made a particular favorite of man. . . . For we see that in its destructive operations—plague, famine, flood, cold, attacks from animals great and small, and all such things— it has as little spared him as any other animal. . . . Besides all this, the discord of inner *natural tendencies* betrays him into further misfortunes . . . through oppressions of lordly power, the barbarism of wars, and the like. . . . Man, therefore, is ever but a link in the chain of physical ends. . . . As the single being upon the earth that possesses understanding, and, consequently, a capacity for setting before himself ends of his deliberate choice, he is certainly titular lord of nature, and, supposing we regard nature as a teleological system, he is born to be its ultimate end. But this is always on the terms that he has the intelligence and the will to give to it and to himself such a reference to [final] ends as can be self-sufficing independently of nature. . . . Such an end, however, must not be sought in nature.[56]

The important point here is the separation that Kant effected—in order to put forward his theory of human freedom—between the animal and the moral lives of the human. He assumed that human beings' animal life was given, constant, and was to be provided for by the planet (the *biosphere*, in today's terms). Human history and thinking were concerned mostly with the constant struggle of humans to meet their moral destiny of a "perfect" and just sociability: "nature has given man two different capacities for two different ends, namely, an end for man as animal species and another end for man as moral species."[57]

The Entangled Moral and Animal Lives of Humans

The pressure that "the animal life" of the human species—our material and demographic flourishing (in spite of the gross inequities of human societies)—now puts on the distribution of natural, reproductive life on Earth, endangering human existence in turn, is something that becomes clearer by the day. It is not surprising then that thinkers and philosophers should call climate change the greatest ethical challenge of the day and raise some critical moral-theological questions, revisiting, in secular forms, the Biblical proposition of "man's dominion over earth": What should humans do, now that our animal/natural life overwhelms the natural lives of nonhumans? Indeed, the question of capitalism reemerges in this morally charged context. Should we continue with capitalism but without fossil fuels? Should we be seeking alternatives to capitalism? Should humans retreat back into small communities? Should the wealthy consume less?

These moral questions testify to the endurance of one of Kant's propositions: that the moral life of humans assumes that man can "choose his way of life and not be bound like other animals to only a single one."[58] But if what I have argued above is right, then it could also be said that the Kantian fable of human history that I recounted is now coming under strain in unprecedented ways. On the one hand, many thinkers still work with (implicitly Kantian) ideas about our moral life representing a zone of freedom, but we cannot any longer afford the assumption that Kant along with many others made—that the needs of our animal life will be attended to by the planet itself. We now want our moral life to take charge of our natural life, if not of the natural lives of all nonhumans as well. The Biblical question of man's dominion has now assumed the shape of secular questions about man's stewardship of and responsibility to the planet.[59]

For reasons of space, let me work with only two prominent examples here of such thinking: Pope Francis's recent and prominent encyclical to Catholic bishops, and a recent essay by Amartya Sen. The pope's encyclical is probably the only available Western/European attempt so far to read humanity's current climate crisis in terms of a deep-set spiritual crisis of modern civilization, albeit within the terms of Catholic theology, but that does not lessen its value. (For an Indian scholar, it is reminiscent of a famous essay Rabindranath Tagore wrote in 1941, the year he died, entitled "The Crisis of Civilization.") The pope has quite a radical critique of the excesses of consumerist capitalism and especially of what he sees as "misguided," "tyrannical," "excessive," and "modern" anthropocentrism of "throwaway" civilization that capitalism has spawned and promoted.[60] In this context, he revisits the question of man's "dominion": "An inadequate presentation of Christian anthropology gave rise to a wrong understanding of the relationship between human beings and the world. Often, what was handed on was a Promethean vision of mastery over the world, which gave the impression that the protection of nature was something that only the faint-hearted cared about. Instead, our 'dominion' over the universe should be understood more properly in the sense of responsible stewardship."[61] "We are not God," writes Pope Francis elsewhere in the book, opposing strongly and by implication the view that humans are now the God species. "The responsibility for God's earth means that human beings, endowed with intelligence, must respect the laws of nature and the delicate equilibria existing between the creatures of the world."[62]

Amartya Sen makes a similar argument but within a non-Christian framework drawing on some tenets of Buddhist thought. Writing on the climate crisis and on human responsibility to other species, Sen argues for the need for a normative framework in the debate on climate change, one that he thinks—and I agree—should recognize the growing need for energy consumption by humans if the masses of Africa, Asia, and Latin America are going to enjoy the fruits of human civilization and acquire the capabilities needed for making truly democratic choices. But Sen also recognizes that human flourishing can come at some significant cost to other species and therefore advocates a form of human responsibility toward nonhumans. Here is how his argument goes.

> Consider our responsibilities toward the species that are threatened with destruction. We may attach importance to the preservation of these species not merely because the presence of these species in the world may sometimes enhance our own living standards. . . . This is where Gautama Buddha's argument, presented in *Sutta Nipata*, becomes directly and immediately relevant. He argued that the mother has responsibility toward her child not merely because she had generated her, but also because she can do many things for the child that the child cannot itself do. . . . In the environmental context it can be argued

that since we are enormously more powerful than other species, . . . [this can be a ground for our] taking fiduciary responsibility for other creatures on whose lives we can have a powerful influence.[63]

There is, of course, some irony in the fact that one of the species "threatened with [at least partial] destruction" is the human species itself. Humans need to be responsible to themselves, which, as the history of humanity shows, is easier said than done. But think of the problems that follow from this anthropocentric placing of humans in loco parentis with regard to "creatures on whose lives we can have a powerful influence." We never know of all the species on which our actions have a powerful influence; often we find out only with hindsight. Peter Sale, the Canadian ecologist, writes, for example, about "all those species that may be able to provide goods [for humans] but have yet to be discovered and exploited, and those that provide services of which we simply are unaware."[64]

This applies even more to the life-form that constitutes the "sheer bulk of the Earth's biomass": microbial life (bacteria and viruses). As Martin J. Blaser observes in his book *Missing Microbes*, microbes not only "outnumber all the mice, whales, humans, birds, insects, worms, and trees combined—indeed all the visible life-forms we are familiar with on Earth—they . . . outweigh them as well."[65] Could we ever be in a position to value the existence of viruses and bacteria hostile to us except in so far as they influence—negatively or positively—our lives? Here again the question is complicated by the fact that ecology and pathology often give us changing and contrary perspectives. Bacteria and viruses have played critical and often positive roles in human evolution, such as the ancient stomach bacteria *Helicobacter pylori*. But since the rise of antibiotics and the consequent changes in the biotic environments of our stomachs, however, *H. pylori* has come to be seen as a pathogen.[66] We cannot be responsible stewards for these life-forms even when we cognitively know about the critical role they have played—and will continue to play—in the natural history of life, including that of human life itself.[67]

This would mean that humans could only ever discharge the responsibility Sen tasks them with imperfectly, since they would never fully know who exactly their wards were or for whom they could assume responsibility in a fiduciary sense. But here indeed is evidence of the strain under which the Kantian fable of human history currently labors. Kant did not demand of human morality that it brought within its own conspectus the natural history of life. Needless to say, his framework was based on a pre-Darwinian understanding of the history of natural reproductive life and constructed long before humans began to discover and understand the roles of microbes in the history of life. We are at a point, however, where we are debating the question of extending the sphere of human morality and justice to include the domain of natural reproductive life.

It is, of course, undeniable that questions of justice between humans have been central to the tradition of the postwar humanities. The intensification, globally, of capitalist forms of social organization has sharpened the political instincts of scholars in the human sciences. Furthermore, given the history of human values in the second half of the twentieth century, we are committed in principle to securing the life of every human and to ensuring their moral and economic flourishing regardless of the overall size of the human population and its implications for the biosphere.[68] Besides, any practical proposal for reducing the size of the human population in effect becomes an antipoor proposition and is therefore morally repugnant. At the same time, a single-minded focus on human welfare and intrahuman justice increasingly seems inadequate. This is the dilemma to which thinkers in the humanities who ponder questions of modernity need to respond. The question is, Since what the humanities and the human sciences provide are perspectives from which to debate the issues of our times, can they overcome their hallowed and deeply set anthropocentrism and learn to look at the human world also from nonhuman points of view?

To Latour, Looking Ahead

Bruno Latour developed his art of thinking long before many of us woke up to the problem he was responding to: the problem posed to modern thought by the unsustainable opposition between nature and science on the one hand and culture and society on the other. He has developed his thinking over the number of texts including the recent *An Inquiry into Modes of Existence*.[69] Since I have been discussing microbial life in this essay, however, let me turn to the classic book of his that speaks of microbes, *The Pasteurization of France*, to show how his thinking clears a path for developing an approach that challenges human modes of being and knowing and helps us to see where the human receives intimations of the nonanthropocentric precisely through the rustle of language that no doubt remains, ultimately, all too human.[70] Additionally, it remains a nice coincidence for this chapter that Latour's anticolonial humor in his book is aimed in part at the good old philosopher from Königsberg whose titanic presence in all discussions of modernity, and for all the barbs we can throw at him, is impossible to escape.

Quite early on in his study of Pasteur's work, Latour draws our attention to the agential presence of microbes not only within the constrained conditions of the laboratory but in everyday human life. "A salesman sends a perfectly clear beer to a customer," writes Latour, but "it arrives corrupted." Why? Because "between the beer and the brewer there was something that sometimes acted and sometimes did not. A *tertium quid*: 'a yeast,' said the revealer of microbes."[71] The presence of microbes tells Latour that "we cannot form society with the social alone": we have to add in "the action of microbes."[72] Thus, "you organize a demonstration of Eskimos in the museum. They go out to meet the public, but they *also* meet cholera and die. This is very annoying, because all you wanted to do was to show them and not to kill them." "Traveling," similarly, "with cow's milk is another animal that is not domesticated, the tuberculose bacillus, and it slips in with your wish to feed your child. Its aims are so different from yours that your child dies."[73] Thus, it is only after the milk has undergone the process of Pasteurization—and the project of purification that commodification entails (chap. 4)—and the microbe has been "extirpated" that it will come to represent the purely "social," that is, "economic and social relationships in the strict sense," which can only happen in some very limited and technologically produced conditions.[74] Latour concludes the first part of his book by remarking that "as soon as we stop reducing the sciences to a few authorities that stand in place of them, what reappears is not only the crowds of human beings, . . . but also the 'nonhuman.'"[75] His project becomes that of "the emancipation of the nonhumans" from what he calls "the double domination of society and science."[76]

Microbes speak of deep time in the history of life. "For about 3 billion years," writes Blaser, "bacteria were the sole living inhabitants on Earth. They occupied every tranche of land, air, and water, driving chemical reactions that created the biosphere, and set conditions for the evolution of multicellular life."[77] Emancipating such nonhumans from the "double domination of science and society" could not be a political task in any institutional sense of the political. Nor does it produce an immediate program of activism. It is a question, primarily and at the current state of development of the governing institutions of humans, of developing a nonanthopocentric perspective on the human world.

In the second part of the book, "Irreductions," Latour looks on this project of "emancipation" of the nonhuman as something akin to an intellectual act of decolonization. "Things-in-themselves?" he puts this rhetorical question to Kant with his characteristic wit, and retorts, "But they're fine, thank you very much. And how are you? You complain about things that have not been honored by your vision?" Latour's critique of the anthropocentrism of Kant's thinking uses the metaphor-concept of colonization to create agential space for the nonhuman. "Things in themselves lack nothing, just as Africa did not lack whites before their arrival," he writes. "However, it is possible to force those who did perfectly well without you to come to regret that you are not there. Once things are reduced to nothing, they beg you to be conscious of them and ask you to colonize them." And he proceeds to place Kant in a line of colonial heroes: "You are the Zorros, the Tarzans, the Kants, the guardians of the widowed, and the protectors of orphaned things."[78] "What would happen," he asks further, "if we were to assume instead that things left to themselves are lacking nothing?"

This is also where the idea of deep time becomes a part of his critique: "For instance, what about this tree that others call *Wellingtonia*? . . . If it is lacking anything, then it is most unlikely to be you. You who cut down woods are not the god of trees. . . . It is older than you. . . . Soon you may have no more fuel for your saw. Then the tree with its carboniferous allies may be able to sap *your* strength." And he drives home the limitations of calculating on human timescales alone (which is what we do when we think politically): "So far it [the tree] has neither lost nor won, for each defines the game and time span in which its gain or loss is to be measured."[79]

And then comes the arrow of a question aimed as much at the heart of ancient Biblical thought as at its Heideggerian mutation, one that declared humans to be specially destined to exercise dominion over the planet:

> Who told you man was the shepherd of being? Many forces would like to be shepherd and to guide others as they flock to their folds to be sheared and dipped. There are too many of us, and we are too indecisive to join together into a single consciousness strong enough to silence all the other actors. Since you silence the things that you speak of, why don't you let them talk by themselves about what is on their minds? Do you enjoy the double misery of Prometheus so much?[80]

This I regard as the most important civilizational question of our times, the one that the pope raised within the limits of his religion.

Latour's epochal question reminds us that deep pasts and futures are not amenable to human-centered political thought or action. This does not mean that our usual disputations about intrahuman (in)justice, inequalities, and oppressive relationships will not continue; they will. But now that the moral and biological lives of the species *Homo sapiens* cannot any longer be disentangled from each other, one has to learn to have recourse to forms of thought that go beyond—but that do not discard—the human political. The connected stories of the evolution of this planet, its climate, and its life cannot be told from any anthropocentric perspective. These other stories are necessarily anchored in accounts of deep time. They make us aware that humans come very late in the history of this planet, that the planet was never engaged in readying itself for our arrival, and that we do not represent any point of culmination in the planet's story. This is where Latour's—and some other scholars'—attempt to open up vistas of aesthetic, philosophical, and ethical thought help us to develop points of view that seek to place the current constellation of environmental crises in the larger context of the deeper history of natural reproductive life on this planet. This returns us to our discussion of planetarity that we take up in the concluding section of the book.

PART III

Facing the Planetary

7
Anthropocene Time

Many Anthropocenes?

The Anthropocene is the perhaps the only term of geological periodization that has been widely debated among humanist scholars with no formal training in stratigraphy, the branch of geology concerned with the ordering of earthly strata and their relationship to geological time. "There are many Anthropocenes out there, used for different purposes along different lines of logic in different disciplines," writes the earth scientist Jan Zalasiewicz.[1] The different Anthropocenes Zalasiewicz mentions circulate in the human sciences as partisan, passionate accounts of what caused the Anthropocene, from when it should be dated, who is responsible for the onset of this epoch, and even what the proper designation of this epoch should be. Many argue about the politics of the name and propose, for instance, that the epoch be more properly called "the Capitalocene" or "econocene" so that a vague and undifferentiated humanity—"anthropos"—is not held responsible for bringing about this uncertain time and that the blame is laid squarely at the door of a system: capitalism or the global economic system.

The Anthropocene debate thus entails a constant conceptual traffic between earth history and world history. There is widespread recognition now that we are passing through a unique phase of human history when, for the first time ever, we consciously connect events that happen on vast geological scales—such as changes to the whole climate system of the planet—with what we might do in the everyday lives of individuals, collectivities, institutions, and nations (such as burning fossil fuels). There is also agreement—however provisional—among scholars who debate the term *Anthropocene* that, irrespective of from when we date it (the invention of agriculture, expansion and colonization by Europe, the Industrial Revolution, or the first testing of the atomic bomb), we are already *in* the Anthropocene.

The Anthropocene requires us to think on the two vastly different scales of time that earth history and world history respectively involve: the tens of millions of years that a geological epoch usually encompasses (the Holocene seems to have been a particularly short epoch if the Anthropocene thesis is right) versus the five hundred years at most that can be said to constitute the history of capitalism. Yet in most discussions of the Anthropocene, questions of geological time fall out of view, and the time of human world history comes to predominate. This one-sided conversion of earth-historical time into the time of world history extracts an intellectual price, for if we do not take into account earth history processes that out-scale our very human sense of time, we do not quite see the depth of the predicament that confronts humans today. Zalasiewicz's arresting remark that to link the problem of the stratigraphic boundary separating the Anthropocene from its predecessor epoch, the Holocene, with events in the world history of humans alone "would run counter to a peculiarity of geological time, which is that, at heart, it is *simply time*—albeit in very large amounts" serves as my point of entry into the Anthropocene debate.² This chapter develops a distinction that Zalasiewicz introduces in this context between human-centered and planet-centered thinking.

But before we follow up on the logic of Zalasiewicz's argument that brings into view the geological aspect of the time of the Anthropocene, we need to begin by explaining why the time of geology presently flits in and out of our attention.

Why Geological Time Falls Out of the Anthropocene Debate

Even though it refers to a new period in the planet's geological history and therefore to geological time, the term *Anthropocene* was used from its very inception as a measure not of geological time—it is not a unit of time—but of the *extent* of human impact on the planet. According to John Bellamy Foster, the appearance of the Soviet geochemist Vladimir I. Vernadsky's pioneering book *The Biosphere* in 1926 "corresponded to the first introduction of the term Anthropocene (together with Anthropogene) by his colleague, the Soviet geologist Aleksei Pavlov."³ From the very beginning, the term referred to the extraordinary scale of human influence on the planet. Foster cites Vernadsky on the subject: "Proceeding from the notion of the geological role of man, the geologist A. P. Pavlov [1854–1929] . . . used to speak of the *anthropogenic era*, in which we live. . . . He rightfully emphasized that man . . . is becoming a mighty and ever-growing geological force."⁴

The recent revival of the term originates from a conference of Earth system scientists in Mexico, where the renowned chemist Paul Crutzen is said to have angrily remarked, "Stop using the word Holocene. We're not in the Holocene any more. We are in the . . . the . . . the . . . the Anthropocene!"[5] When later, in the year 2000, Crutzen and the lake biologist Eugene F. Stoermer proposed a general adoption of the idea of the Anthropocene, it was not the problem of time that was foremost in their considerations. They saw the word as a convenient shorthand for pointing to the size of the human footprint on the planet: "Considering . . . [the major] and still growing impacts of human activities on earth and the atmosphere . . . at all, including global, scales," they recommended the term *Anthropocene* for "the current geological epoch" as a way of registering "the central role of mankind in geology and ecology."[6]

The term *Anthropocene* helped focus public attention on the possibility that human beings now so dominated the planet that their collective impact was comparable to those of very large-scale planetary forces. The paleoclimatologist David Archer clearly saw the term *Anthropocene* as a rough measure of human impact on earth processes: "Geologic time periods in the past are generally delineated by major changes in climate or by biological extinctions. Earth's alleged graduation from the Holocene to the Anthropocene is therefore a statement that humankind has become a powerful force in Earth evolution."[7] He even gave us a precise estimate of the kind of planetary geophysical force that humans had become:

> The deepest and most profound climate changes seem to take place on timescales of millennia and longer. The great ice sheets grow and usually melt on timescales of millennia, a huge response to wobbles in the Earth's orbit. The natural carbon cycle acted as a positive feedback, amplifying the response to the orbit. . . . Human climate forcing has the potential to overwhelm the orbital climate forcing, taking control of the ice ages. Mankind is becoming a force in climate comparable to the orbital variations that drive the glacial cycles.[8]

In explaining the term *Anthropocene* in 2011, a good ten years after it had been proposed, Will Steffen, Jacques Grinevald, Paul Crutzen, and John McNeill reiterated that the "concept of the *Anthropocene* . . . was introduced to capture this quantitative shift in the relationship between humans and the global environment. . . . Humankind . . . rivals some of the great forces of Nature in its impact on the functioning of the Earth system" and has become "a global geological force in its own right."[9] Talking about a new geological epoch was a way of emphasizing the sheer scale of human impact on the planet.

Discussions, scientific or not, of human impact on the planet's environment could never be completely separated from moral concerns.

Should humans have so large an impact at all? Could they even afford to have such an impact without imperiling their own existence? These and similar questions were never far from the concerns of the researchers mentioned above. This is why they took on the citizenly role of publicizing their findings. Such moral concerns have perhaps always accompanied attempts to quantify human impact on Earth. It frames, for instance, John R. McNeill's landmark book *Something New Under the Sun: An Environmental History of the Twentieth-Century World*, published in 2000, perhaps the most remarkable attempt to date by a historian to document meticulously human impact on the resources, atmosphere, and the biosphere of the planet. The book is framed by a moral judgment McNeill makes at the beginning: "Albert Einstein famously refused to 'believe that God plays dice with the world.' But in the twentieth century, humankind has begun to play dice with the planet, without knowing all the rules of the game."[10] Even the authors of a pioneering 1957 scientific paper on "Increase of Atmospheric CO_2 during the Past Decades" (now considered to have been of historical importance in the development of the science of anthropogenic climate change), Roger Revelle and Hans E. Suess, could not help using words that clearly reached beyond the purely scientific. "Human beings," they said, "are now carrying out a large-scale geophysical experiment of a kind that could not have happened in the past nor be reproduced in the future. Within a few centuries we are returning to the atmosphere and oceans the concentrated organic carbon stored in sedimentary rocks over hundreds of millions of years."[11] A rising sense of alarm as climate science progressed in the 1970s and 1980s resulted in the establishment in 1989 of the Intergovernmental Panel on Climate Change (IPCC). What was still a grand "experiment" on the part of humanity in Revelle and Suess's prose in 1957, as the IPCC presented their various assessment reports through the 1990s and the 2000s, transformed into a message warning governments about the risks of a "dangerous" climate change facing humanity.

From the very beginning of its career, then, the Anthropocene has had two lives, sometimes in the same texts: a scientific life involving measurements and debates among qualified scientists, and a more popular life as a moral-political issue. So long as the Anthropocene was seen mainly as a measure of human impact, though acknowledged as the impact that ushered in a new period in the planet's history, the focus remained on the force and its wielder (humanity, capitalist classes, rich nations, capitalism), and questions of geological time simply fell into the shadows. Moral questions about culpability and responsibility have necessarily dominated this debate. Not surprising, perhaps, if we remember

Sheila Jasanoff's observation that "representations of the natural world attain stability and persuasive power . . . not through forcible detachment from context, but through constant, mutually sustaining interactions between our senses of the *is* and the *ought*: of how things are and how they should be."[12]

Translating "Force" into "Power," from Earth History to World History

It is the moral side of the Anthropocene debate—questions of historical responsibility for the warming that has happened so far—that requires us to translate ideas that have deeply to do with earth history, geology, and geological time into the language of world/human history.[13] This entails, however, two important acts of displacement: the displacement-translation of the category "force"—referring to the physical pull that one material body exerts on another (to go by the Newtonian understanding of it), thus humanity as a geological force—into the human-existential category of power and its sociological-institutional correlates, and the accompanying dislodging of the problem of the Anthropocene from the realm of geological time to the time of human or world history.[14]

The displacement of the category of physical force onto the historical-existential category of power is visible in the writings of two groups of scholars and activists: those who want to assign culpability for the offense of creating the global environmental crisis, and those who seek in the crisis of global warming an ethical horizon for the future of humanity as a whole. Sometimes, we may find both tendencies in the same text. Take two documents that were published in the early 1990s: the first-ever report of the IPCC that was published in 1990, and the 1991 tract *Global Warming in an Unequal World*, by two Indian environmental activists, the late Anil Agarwal and Sunita Narain, whom we encountered before (e.g., in chap. 4). "There is *concern*," said the first IPCC report in its summary for policy makers, "that human activities may be inadvertently changing the climate of the globe through the enhanced greenhouse effect . . . which will cause the temperature of the Earth's surface to increase. . . . If this occurs, consequent changes may have a significant impact on society."[15] Agarwal and Narain objected to such sweeping use of the word "human," though the immediate target of their polemic was not the first IPCC report but a report of the World Resources Institute (WRI) on the "global environment" published in the same year as the IPCC report (1990).[16] It was this report, readers will remember, that Agarwal and Narain described as an "excellent example of environmental colonialism."[17]

For Agarwal and Narain, as we have noted before, it was as though the talk about climate change was creating a cruel and unfair "regime of historicity"—to speak with François Hartog—that threatened to foreclose the world-historical time of development in which the future was an open vista of modernization that the United States and Soviet Union inspired after the Second World War.[18] They could not see what today may look like an Asian future for the crisis of climate change. They shared the "fear" of "many developing countries" that the climate talk was aimed to "put serious brakes on their development by limiting their ability to produce energy, particularly from coal."[19] This is why one could see the argument for "climate justice" as an argument in world history (which is not to deny the developing nations' point about climate justice). A particular and familiar narrative of European imperialism was encrypted in the reference to "colonialism" in the very title of Agarwal and Narain's tract and the explicit third-worldist vocabulary of their text. This was putting the problem squarely in terms of world history.

Once the idea of the Anthropocene had been mooted, Swedish academics Andreas Malm and Alf Hornborg were among the first to fire a salvo against the proposition that global warming was "anthropogenic" in nature, objecting, in the manner of Agarwal and Narain, to the use of the word *anthropos*. "Realising that climate change is 'anthropogenic,'" they wrote, "is really to appreciate that it is *sociogenic*."[20]

> The succession of energy technologies following steam—electricity, the internal combustion engine, the petroleum complex: cars, tankers, aviation—have all been introduced through investment decisions, sometimes with crucial inputs from certain governments but rarely through democratic deliberation. The privilege of instigating new rounds appears to have stayed with the class ruling commodity production.

Citing the facts that "as of 2008, the advanced capitalist countries or the 'North' composed 18.8% of the world population, but were responsible for 72.7 [per cent] of the CO_2 emitted since 1850," they asked: "Are these facts reconcilable with a view of *humankind* as the new geological agent?" Starting from the premise that "uneven distribution is a condition for the *very existence* of the modern, fossil-fuel technology," they argued for the "need to probe the depths of social history," something that "geologists, meteorologists and their colleagues are not necessarily well-equipped to study."[21] The need of the hour was to stay faithful to—and not "abandon"—"the fundamental concerns of social science, which importantly include theorization of *culture and power*."[22] How else, they asked in concluding their essay, "can we even imagine a dismantling of the fossil[-fuel] economy?" "Species-thinking on climate change is conducive to mystification and political paralysis."[23]

Many others have followed suit, notably among them the sociologist Jason Moore, who recommended that the new geological epoch be given a name suggestive of the more immediate factors that in his opinion brought it about: Capitalocene.[24] Moore acknowledged that this "is an ugly word in an ugly system," but "the Age of Capitalism does not merit an aesthetically pleasing moniker."[25] I do not wish to either endorse or criticize the appellation on which Moore decided; my point is to show how applying this nomenclature entailed, once again, the act of folding the concept of "force"—humans as a geophysical force—into the human-existential category of power that is intrinsic to world history. Here is Moore on the subject, beginning with a bit of banter: "[The dominant Anthropocene narrative] tells us that the origins of the modern world are to be found in England. . . . The motive force behind this epochal shift? In two words: coal and steam. The driving force behind coal and steam? Not class. Not capital. Not imperialism. Not even culture . . . you guessed it: the *Anthropos*: Humanity as an undifferentiated whole." And his critique follows: "The Anthropocene makes for an easy story . . . because it does not challenge the naturalized inequalities, alienation, and violence inscribed in modernity's strategic relations of *power* and *production*. . . . This erasure, this elevation of the *Anthropos* as a collective actor has encouraged . . . a meta-theory of humanity as a collective agent, without acknowledging the forces of capital and empire that have cohered in *modern world history*."[26] Needless to say, the word *force* used here by Moore with reference to capital does not connote the Newtonian meaning of the word.

Ian Angus, who has produced a thoughtful Marxist-historical analysis of the Anthropocene—and who, incidentally, does not like the term *Capitalocene* and acknowledges that Earth system scientists recommending the Anthropocene do not necessarily deny questions of climate justice or human differentiation—effects the same displacement by splitting the Anthropocene into two separate phenomena: a "biophysical" Anthropocene and a "socioeconomic" one.[27] The biophysical Anthropocene—"a qualitative change in Earth's most critical physical characteristics that has profound implications for all living things"—is important, "but to properly understand the Anthropocene, we must see it as a *socioecological* phenomenon," the "culmination of two centuries of capitalist development," a period of "economic and social change during which the Holocene ended and the Anthropocene began."[28]

The displacement-translation of "force" into "power" is also undertaken by those who, in order to motivate humans to do something to mitigate the effects of their planetary footprint, appeal to the human sense of their own timescales. Even Earth system scientists defending the idea of geological time have found it important, strategically, to concede the point that "in wider society, geological time-scales are often used as reasons for non-action on societal, intergenerational, and individual timescales ('climate has always changed,' 'coral reefs have become extinct several times, but reappeared' and so on)."[29]

The sense that the scientist-communicator of anthropogenic climate change has to constantly move between different scales of time haunts, for example, Archer's book *The Long Thaw*. Archer's geologist eyes are trained to see how humans have already changed the climate of the planet for the next one hundred thousand years at least. But, he asks, in the very first chapter of the book, "Why should we mere mortals care about altering climate 100,000 years from now? . . . The rules of economics, which govern much of our behavior, tend to limit our focus to even shorter time frames." So Archer uses temporal scales that can connect to the reader's sense of pride and shame. "How would it feel," he asks his reader, "if the ancient Greeks, for example, had taken advantage of some lucrative business opportunity for a few centuries, aware of potential costs—such as, say, a stormier world, or the loss of 10% of agricultural production to rising sea levels—that could persist to this day? This is not how I want to be remembered."[30] This may not be effective rhetoric goading people into action, but the translation of physical "force" into the very human terms of "power" and "responsibility" may be seen to be at work in all texts searching for a planetary human ethics in the present time.

Both geological time and historical time are expressive of human categories, but they are tinged with different kinds of affect. It is, of course, only within the sense of time that informs world history that we can speak of hope or despair. A certain degree of metaphorical use of the idea of the Anthropocene is therefore recommended by some Earth system scientists themselves, and note their quick switch from "force" to "power." "The Anthropocene used as a metaphor might help trigger new normative and ethical thinking. If humanity now has the *power* of being a 'geological force,' it follows that such power should be used carefully and sparingly. . . . That, at least, might enable the Anthropocene to symbolize hope rather than despair."[31] This, of course, assumes that humanity is one and that this "one" can act as an individual person does, using one's capacity ("power to be a geophysical force") with care and responsibility. The astrobiologist David Grinspoon's recent book with the telltale title *Earth in Human Hands* gives us yet another example of what I have come to think of as code-switching between the physical category of force and the social-existential categories of "consciousness" and "power." "Nobody can credibly deny," he writes, "that we are in a time of rampant human influence on Earth. Defined in this crude way, the Anthropocene obviously exists, so why insist it must be bad? What do you propose? That we convince everyone to feel bad about their rotten species?" The ethical task, Greenspoon thinks, is for humanity to be a *conscious* geological force: "Our choice is over what kind of human-influenced Earth we will have. We may lament this truth, but we no longer have the option to choose not to be geological change agents. . . . How to do it right—*that* should be our concern."[32] Humans, writes the earth scientist Daniel Schrag, are at a "point of no return." "In the Anthropocene," he adds, " the survival of nature as we know it may depend on the control of nature [by humans]—a precarious position for the future of society, of biological diversity and of the geobiological circuitry that underpins the Earth system."[33] Clive Hamilton, who has played a pioneering role in discussions of climate change by humanist scholars, argues in his book *Defiant Earth* for a "new anthropocentrism"—likening humanity to a "conscious force." In a "geological epoch in which humans now rival the great forces of nature," the "future of the entire planet, including many forms of life, is now contingent on the decisions of *a conscious force*, even if the signs of it acting in concert are only embryonic (and may be still born). In the face of this brute fact, . . . denying the uniqueness and power of humans becomes perverse."[34]

If we had to name, from among world historians, a patron saint for this vision of a world-historical future for humanity in which humans take collective responsibility for their physical impact on the planet, it would be William H. McNeill. At a world history conference held in 1994 at Wesleyan University, he even proposed a world-historical role for world historians: "by constructing a perspicacious and accurate world history, historians can play a modest but useful part in facilitating a tolerable future for humanity as a whole and for all its different parts . . . inasmuch as a clear and vivid sense of the whole human past can help to soften future conflicts by making clear what we all share."[35] This called for an intellectual partnership between scientists and world historians, as McNeill argued a few years later in a 2002 essay: "It is time for historians to . . . begin to connect their own professional thinking and writing with the revised scientific version of the nature of things."[36] A total history of humanity was the history of the species: "We are . . . at one with our predecessors, immersed in processes we do not control and can only dimly understand—a process nonetheless that has made us . . . the most disturbing . . . and . . . extraordinarily powerful factor in upsetting the multiple levels of . . . equilibria within which we exist. . . . Perspicacious history of how we got to where we are might even improve human chances of survival."[37] A year later, in 2003, he wrote, "our species as a whole [has become] an unexampled threat to other life-forms. Long-term disaster may well loom ahead: but so far so good . . . humankind's greatest age may still lie ahead. Or, just as probably, we may be precipitously rushing toward any of several disastrous terminations of our altogether extraordinary career on earth."[38]

This turn toward human capacities for a solution to our global environmental crisis also marks the end of John L. Brooke's magisterial survey of the history of humanity through various climate regimes on this planet. "In the final analysis," writes Brooke,

> our current circumstance needs to be seen both as a crisis in the relation of humanity to the earth system, and as a moment in the long-term transformation of economic systems on a scale with any of the great ruptures of the human past. . . . What is necessary, what all the pragmatists are working for, what the pessimists despair of, and what the deniers reject . . . is a global solution. *We hold it in our collective capacity to address the earth system crisis that is now upon us*. That capacity must be mobilized by an informed political will.[39]

Once, again, the solution to problems on the scale of earth history is sought in the human timescales of politics and world history. I will have more to say about the displacement effected here.

The Time of World History

The time of world history is, ultimately, the same as what Reinhart Koselleck identified as the time of human history. The texture of human-historical time, as Koselleck famously suggested, is made up of the warp and woof of two fundamental categories that for Koselleck constituted "an anthropological condition" for history itself: "the space of experience" and "the horizon of expectation."[40] Koselleck expressed powerfully what many thinkers over the ages had thought about the human sense of historical time. Recall Augustine, for instance: "The present of past things is memory; the present of present things is direct perception; and the present of future things is expectation."[41]

Neither human-historical time nor the time of geology, both being of human making, is empty of affect. But they engage, as mentioned before, very different types of affect. There have, of course, been arguments about whether or not the sheer chronology of world history should be looked upon as working like a sack of empty time indifferent to the events we pour into it. Some scholars have recommended this thought on moral grounds:

> Empty time has to be taken in a sense that implies more than a mathematical method to bring abstract order to given data. Time has to be taken as a potential bond of life, history as a garden with a common concept of life, real life. This is the only way to provide a common ground for historical narratives, for keeping history as a universal reality together. We may produce all kinds of historical concepts and historical temporality, but we do not escape the necessity to hold fast to the concept of empty time as the open field on which histories may arise, keeping in touch with one another.[42]

Koselleck's anticipatory rebuttal of this point is also worth recalling. He agrees that in constructing historical time that is always "tied to social and political units of action, to particular acting and suffering human beings, and to their institutions and organizations," one may very well need "measures of time that derive from the mathematical-physical understanding of nature . . . : the dates or length of a life or of an institution, the nodal or turning points of political or military series of events, . . . [and so on]." But such a mathematical-physical understanding of time cannot act as the ground for human history: "an interpretation of the interrelations that result already leads beyond the natural or astronomically processed determinations of time. Political constraints on decisions made . . . [and other considerations], in their mutual interaction or dependence[,] finally [force] us to adopt social and political determinations of time that, although they are naturally caused, must be defined as specifically historical."[43]

Experience, Koselleck explains, is "present past" and could include a "rational reworking" of the past as well as "unconscious modes of conduct which do not have to be present in awareness." Expectation is "the future made present" oriented to "that which is to be revealed."[44] The two could interpenetrate—"only the unexpected has the power to surprise and this surprise involves a new experience"—and Koselleck would spend many pages explaining how in the time of the modern, in *Neuzeit*, "the difference between experience and expectation is increasingly enlarged," and "eager expectations" may also "remove themselves from all previous experience."[45] "In brief," he summed up, "it is the tension between experience and expectation that, in ever-changing patterns, brings about new resolutions and through this generates historical time."[46] This means historical time cannot be separated from certain kinds of human affect: "prospects of the future, raising hopes and anxieties, making one precautionary or planful," they all go into the making of historical time.⁴⁷ This is what climate change as "world history" is: a stage for the play of various human emotions including those of hope and despair. One could indeed look upon the Paris climate deal of 2015 as such an intense and frenetic piece of world history.⁴⁸

In contrast, one could say that the human affect that usually relates to the time of geology would be very different. Several geological periods, personalities, and events have, of course, entered human time as culturally processed phenomena—the Jurassic age of dinosaurs, for instance, or the 1815 eruption of the Tambora volcano in Indonesia.⁴⁹ But most geological events do not undergo such affective processing. We have no obvious emotions about the great oxygenation event of 2.5 billion years ago—though human life would be inconceivable without that event—or about the Ordovician-Silurian great extinction event that took place more than 440 million years ago.

Thinking Geological Time

How then does the question of "simple" geological time—time to which Earth system history, with its million-year carbon cycles, properly belongs—erupt in this landscape of understanding that constantly relocates both the ideas of humans wielding a geophysical force and the new geological epoch of the Anthropocene in the affective past, present, and future of human power and responsibility?

The recent story of the Anthropocene reversed the usual relationship between geologists' work and the big themes of human or even other kinds of histories.⁵⁰ "Geologists tend not to think about history, much," writes Zalasiewicz, for the story they eventually want to put together concerns not only the geology of this planet but also of "the billion strong" planets and moons orbiting "other stars in the galaxy," not to mention "the planetary bodies that will be present in the one hundred billion or so other galaxies within the known Universe" that we cannot even see into yet. So how does a geologist get to place "any particular, strange and novel event within" a big story "such as—for instance—the extraordinary set of processes that we humans have precipitated?" From where does a practicing geologist start to think about the proposed new geological epoch of the Anthropocene?

The usual starting point for the geologist, writes Zalasiewicz, is seldom the big story itself but "fragments"—"small shards of the greater whole that have attracted the attention of some passing geologist, using that last word extremely loosely." The larger synthesis "typically emerges" once sufficient detail has been "collected together" to generate recognizable patterns in "what seemed initially to be bewilderingly chaotic."⁵¹ He gives the example of the Carboniferous period that lasted from about 359 to 299 million years ago and that produced coal-rich strata of rocks. Generations of geologists mapped out these rocks in extreme detail for practical "here and now" purposes. The larger story that "those Carboniferous rocks are a memory of something else entirely—of a world of primeval swamp forests, with amphibians and giant dragonflies, without flowers, or birds, or mammals" was seldom the main concern of working geologists. The big history of that distant past, "now separated off as a segment of time some 60 million years long"—the story of the Carboniferous period—"may now be reconstructed in imagination" from these rocks but "never again touched, or seen, or experienced."[52]

However, so long as we think of the name and the concept of the Anthropocene as a measure—and a critique—of the impact humans have had on the geobiology of the planet, we cannot escape the moral pull of world history, for questions of empires, colonies, institutions, classes, nations, special-interest lobbies—in a word, the world system created by European empires and capitalism—are then never far from our concerns. This is clearly the reason why the Anthropocene, perhaps, is the only suggested name of a geological period that has critically engaged— if not outraged—many scholars in the human sciences. The archaeologist and anthropologist Kathleen D. Morrison, for instance, proposed that the task at hand was to "provincialize the Anthropocene" in order to expose the "hidden Eurocentrism" of the concept. It represented, in her judgment, "an effort to expand (rather homogenized) European historical experiences, frameworks and chronologies onto the rest of the world." For her, the problem remains that "most proposals for an Anthropocene era adopt a rather limited historical perspective, assuming that significant environmental impact began only with the (European, and especially British) Industrial Revolution." "Provincializing the Anthropocene" meant therefore "that we no longer take European agricultural or industrial history as a starting point."[53] Instead, Morrison pointed to other possible beginnings: "large-scale human burnings," for instance, that have for a very long time "reshaped vegetation regimes," or agriculture, "another major means by which our species has reshaped not only vegetation, but also soils, slopes, hydrology, disease environments, the distribution of wild plants and animals and has made possible new configurations of human population."[54] One could add to this list megafauna extinction, rice production, and other big events, including the control and management of fire, suggesting the force of human impact on the planet.

In a significant paper published in 2015, two British geographers, Simon L. Lewis and Mark A. Maslin, starting from the premise "that formal establishment of an Anthropocene epoch would mark a fundamental change in the relationship between humans and the Earth system," suggested two possible dates for when the Anthropocene may have begun: 1610 and 1964. They agreed that to define a geological time unit, "formal criteria must be met." Yet dating the beginning of the Anthropocene also remained for them a necessarily moral-political exercise: "defining an early start date may, in political terms, 'normalize' global environmental change. Meanwhile, agreeing [to] a later start date related to the Industrial Revolution may, for example, be used to assign historical responsibility for carbon dioxide emissions to particular countries or regions during the industrial era." Besides, they added, "the formal definition of the Anthropocene makes scientists arbiters, to an extent, of the human-environment relationship, itself an act with consequences beyond geology. Hence there is more interest in the Anthropocene than other epoch definitions."[55]

In the end, Lewis and Maslin preferred 1610 to 1964 as a point from which to date the Anthropocene. They gave evidence-based scientific reasoning for their preference: a decline in atmospheric CO_2 (7–10 ppm between 1570 and 1620) coinciding with a massive decline in population in the Americas following the arrival of Europeans (from 64 million in 1492 to 6 million "via exposures to diseases..., war, enslavement and famine").[56] But they also mobilized world-historical arguments to justify their choice.

> The choice of either 1610 or 1964 [showing a "distinct peak in radioactivity" from detonation of nuclear bombs] would probably affect the perception of human actions on the environment.... [1610] implies that colonialism, global trade and coal brought about the Anthropocene. Broadly, this highlights social concerns, particularly the *unequal power relationships* between different groups of people, economic growth, the impact of globalized trade, and our current reliance on fossil fuels.... Choosing the bomb spike tells a story of *an elite-driven technological development* that threatens planet-wide destruction.[57]

They saw the Anthropocene as something that brought together earth history and world history: "The impact of the meeting of Old and New World human populations—including the geologically unprecedented homogenization of Earth's biota—may serve to mark the beginning of the Anthropocene.... It represents a major event in world history [as well]."[58]

Lewis and Maslin's point of view has been both vigorously criticized and defended.[59] But the Anthropocene, so long as it is seen as a measure of human impact on the planet, can have only plural beginnings and must remain an informal rather than a formal category of geology, capable of bearing multiple stories about human institutions and morality. The issue cannot be separated from political and moral concerns. Questions of stratigraphic significance—such as, is there enough evidence in the strata of the planet for stratigraphers to be able to argue that the thresholds of the Holocene epoch have been exceeded?—then get written over by varieties of human history, deep and shallow, big and small. Zalasiewicz's paper, which I began with, is of interest in this debate for this very reason: it removes—perhaps for the first time in the decade-old controversy about the Anthropocene—the cobweb (or should I say, the human web) of world-historical time to bring into view what he calls the time, "simply," of geology.

Zalasiewicz makes some crucial moves that should be noted. He recognizes that when "the Anthropocene was born—in a practical sense, at least—with Paul Crutzen's inspired improvisation at a conference in Mexico just fifteen years ago, the usual procedures were turned upside down."[60] The idea came from "the Earth system community, monitoring planetary change in real time." But they were not necessarily stratigraphers. When the Anthropocene Working Group of the Subcommission on Quaternary Research, "a component body of the International Commission of Stratigraphy (the decision-making body that oversees the Geological Time Scale)" was set up, the "first task" of the group was "to see whether there is in effect, a stratal unit on Earth that may be systematically recognized and assigned, *as a material body*, to the Anthropocene Epoch." In geologists' parlance, such a "material time-rock unit, parallel to the 'time' unit, would be termed an Anthropocene series."[61] In this particular essay and elsewhere, Zalasiewicz and his colleagues have gone to great lengths explaining what materials (including technofossils) such a time-rock unit would most likely be made of.[62]

This quest for stratigraphic records proper to the Anthropocene is centered on the question of whether it could be argued that there is enough evidence in the lithosphere and on the surface of the planet to support the proposition that the planet has exited the threshold of the Holocene epoch. The critical questions for stratigraphers are not "how globally important"—in human terms—the new boundary is or "when was the first sign of influence of some major new factor in the Earth system?" a question that understandably concerned many who debated the moral aspects of the idea of the Anthropocene. As Zalasiewicz puts it, " in terms of the definition of a 'stratigraphic Anthropocene,' [at issue is] . . . change to the Earth system rather than a change to the extent to which [we] are recognizing human influence." It is important to be able to show with stratigraphic evidence that "the planetary system is *recognizably* changing." The task of making a formal proposition for the Anthropocene does not necessarily require the stratigrapher to be interested in the whodunit part of the story. The impact on the lithosphere is what matters; the author of the impact is not important. The name Anthropocene carries no special literal or human significance for stratigraphers, for "it just happens to be the activities of the human species that are currently the main perturbing force." "The Anthropocene," writes Zalasiewicz, "would remain just as important geologically, because of the scale of the planetary (and hence stratal) effects, if it had some other cause" than human activities. "Indeed," he remarks, "the concept would then probably be rather easier for humans to comprehend and react to."[63] From this stratigraphic perspective—necessary if one were to formalize the new geological epoch—the Anthropocene, as Zalasiewicz puts it, is "seen as a planet-centred, rather than human-centred phenomenon."[64]

It is his concern with what he calls the "stratigraphic Anthropocene" that enables Zalasiewicz to distill his point about geological time as distinct from the time of human history. "The question of the [epochal] boundary has aroused a good deal of comment," he writes, "not least as regards the protracted and progressive nature of significant human influence on the Earth, ranging from the beginnings of the extinctions of the terrestrial megafauna, starting as long as 50,000 years ago . . . through the development and spread of agriculture beginning some ten thousand years ago, to the origin and spread of urbanization a little later." The resulting "time-transgressive human-altered surface layer," called the archaeosphere, has sometimes been seen as "the most visible reflection" of the Anthropocene. But "this would be a parallel," comments Zalasiewicz, "of *archaeological time terms* such as 'Paleolithic,' 'Bronze Age,' and so on, which are all different ages in different regions, reflecting the cultural state [and we might add, power relations] of the local human populations." One could add "Capitalocene" to this list of human-centered rather than planet-centered definitions of the Anthropocene. And they would all, says Zalasiewicz, "run counter to a peculiarity of geological time that, at heart, is *simply time*—albeit in very large amounts. A time boundary (whether geochronological or chronostratigraphical) is just an interface in time, of no duration whatsoever—it is less than an instant—between one interval of time (which may be millions of years long) and another. It is inherently synchronous within the domain across which it operates, which is that of the home planet."[65]

Human-Centered and Planet-Centered Ways of Thinking

Let us stay for a moment with the distinction that Zalasiewicz made between human-centered and planet-centered ways of thinking. Aspects of the thesis about humans constituting a geological force comparable to the Milankovitch effect that controls the glacial-interglacial cycles entail thinking on temporal scales that are indeed too large for any political-affective apprehension and hence for the making of politics or policy. Some of the earth processes are extremely slow in human terms. As David Archer says, the million-year carbon cycle of the planet is "irrelevant for political considerations of climate change on human time scales," but "ultimately the global warming climate event will last for as long as it takes these slow processes to act."[66] Soil, fossil fuel, and biodiversity are not renewable on human timescales. Past catastrophes, write Charles H. Langmuir and Wally Broecker in their book *How to Build a Habitable Planet*, show that "biodiversity recovers only on timescales of millions of years."[67] These are all events or processes that have been affected by human activity but they act themselves out not on the timescale of world history but on geological scales of time.

Geological time is not identical to absolute mathematical time. There remains a *material* side of time for geologists, for there is no geological time without geological objects. Ultimately, for the purposes of our discussion, this time is written into the strata of the planet. "And indeed it is these strata, with their radionuclides, fly ash, microplastics, supermarket chicken bones and so on that form the core of the argument for the 'geological [stratigraphic] Anthropocene,'" writes Zalasiewicz.[68]

But however we think of geological time—and over a long number of years Christian theology (geology as the Book of Nature), astronomy, physics, evolutionary biology, and other areas of thought have contributed to its history—it belongs in part to a class of time that has always been seen (long before geology) as opposed to the sense or scale of temporality of human history.[69] Saint Augustine saw this kind of time as expressed in numbers "to which we cannot give a name"; Buffon thought of it as time that did not "conform to the limited powers of our intelligence"; Darwin described its "vastness" as "incomprehensible"; self-described geologists in the early nineteenth century came to accept it as something that—in the words of one of their great historians—was "literally beyond human imagination" even if "no quantitative figures could [yet] be attached to it."[70] All these descriptions, of course, do not speak of "empty time," shorn, as such, of human affect. Augustine, Buffon, and Darwin all speak of this time only in its relationship to being human, thus marking it as representing a limit to the time of historicality, as a conceptual-temporal place where "meaning making" of human history—the tension between the horizon of expectation and the horizon of experience—ceases to work.[71]

The narrative of world history has now collided (in our thoughts) with the much longer-term geological history of the planet or—as we now think of it—of the Earth system.[72] Earth System Science that draws on planetary histories represents a later and viable mutation of James Lovelock's Gaia hypothesis that was advanced in the 1960s. Without our being able to see the planet as some kind of a system—a system of "steady state disequilibrium" maintained by an external energy source (the sun) that moves interlocking processes and feedback loops supportive of life over the long run—there would have been no science of planetary climate change and no scientific formulation of the problem either.[73] The history of the publication of Langmuir and Broecker's *How to Build a Habitable Planet* captures something of the youth of Earth System Science as a discipline. Broecker published this book in 1984 under the same title and as its sole author. But that was a time, as Langmuir and Broecker point out in the second edition, "when dark energy and dark matter were not yet discovered, the ocean ridges were barely mapped, hydrothermal vents on the sea floor were barely known, the Antarctic ice core had not been drilled, the 'snowball Earth' hypothesis had not been fully formulated, global warming was not yet an urgent topic, and no extrasolar planet had been discovered." One could say, in a Latourian vein, that it took all these technologies and discoveries for scientists to think into being the "Earth system" as an object of study. In their 2012 revised edition of the book, the authors included "discussion of life, . . . earth history, the rise of oxygen, . . . volcanism and the role of the solid Earth in habitability" in addition to taking "a 'systems' approach to the history and understanding of our planet." "If there is one theme that we hope comes through in the book," they wrote in their preface, "it is of a connected universe in which human beings are an outgrowth and an integral part."[74]

While Earth System Science is central to ideas about planetary climate change and understanding the Anthropocene, the key question, as we saw in chapter 3, driving this interdisciplinary branch of scientific knowledge concerns the history of life on earth and the supportive Earth system processes, all considered on geological if not astronomical scales of time. What makes a planet habitable not just for human life but for complex life in general? Do humans have a necessary place in planetary evolution? Are there others like us out there somewhere?[75] "A critical unknown," to recall the words of Langmuir and Broecker we have already encountered in chapter 3, "is the fraction of a planetary lifetime that a technological civilization exists. Does such a civilization self-destruct in a few hundred years or last millions of years? For such a civilization to last, the species . . . must sustain planetary habitability rather than ravage planetary resources."[76]

The habitability problem, so central to astrobiology and so different from the human-centered idea of sustainability, does not even entail any necessary assumption that humans exist on other planets. In imagining technological civilizations elsewhere, all that astrobiologists need to assume is the existence of what they call SWEIT, or "Species with Energy-Intensive Technology."[77] Astrobiology looks at the earth and other planets from an imaginary floating point in space: "For a technological civilization to persist they would need to correspond with a planet as a natural system."[78] Depending on how a SWEIT acted, a planet could go from "being a 'habitable planet' to an 'inhabited planet' i.e. one that carries intelligence and consciousness of a global scale, for the benefit of the planet and all its life." But there could also be an "abortive and failed mutation," and a planet could regress to an earlier stage of evolution of life, suffer reduced biodiversity, or be even rendered virtually dead.[79]

The essay by Zalasiewicz that I have been discussing evinces a similar view of the planet looked at from outside and as if through a series of time-lapse photographs. After the great oxygenation event 2.5 billion years ago, "the world changed colour, going from the greys and greens of a chemically reducing world to reds, oranges, and browns, as a swathe of oxide and hydroxide minerals appeared."[80] Similarly, Langmuir and Broecker's vision of an "inhabited planet" that has internalized technical intelligence is close to the geologist Peter Haff's proposition regarding there being a technosphere on Earth, a layer he considers analytically distinguishable from the lithosphere, atmosphere, or biosphere, and to study which one has to adopt an extraterrestrial point of view: "Humans have become entrained within the matrix of technology and are now borne along by a supervening dynamics from which they cannot simultaneously escape and survive. . . . Technology is the next biology."[81]

The time of such history is the time of Earth System Science, vast and incomprehensible in terms of the concerns of human history though it is available to our cognitive and affective faculties. In old Althusserian terms, the history of the Earth system is all "process without a subject." In the vocabulary of Bruno Latour, this is a narrative of many dispersed and networked actors, none acting with the sense of internal autonomy with which humanist historians suffuse the word *agency*. Yet in social-science debates about the Anthropocene, geological time gets written over by the human time of world history, and humans emerge as the subject of the drama of the Anthropocene, not just in the writings of scholars in the human sciences but often in those of earth scientists themselves. It is clear why it happens, for the science of Earth system history has been made possible by the same technologies that have also produced, mapped, and measured the deleterious impact on the biosphere of the complex of species and life-forms represented by humans, their dependent or coevolving living entities, and their technology. This species-technology complex has flourished at the expense of many other species and now threatens to push the Earth system into another phase altogether.

Texts, Langmuir and Broecker's included, written by Earth system scientists to communicate the message of the current planetary environmental crisis speak necessarily in two voices. They think simultaneously in two ways, as it were: human centered and planet centered. There is the vast story of life on this planet and the general questions of habitability of a planet, questions to which humans are not central. But there is also the theme of the impact of human activities on the earth. "Human civilization has led to the first global community of a single species, destruction of billions of years of accumulation of resources, a change in atmospheric composition, a fourth planetary energy revolution, and mass extinction. . . . The potential for planetary change is almost as great as that caused by the origin of life or the rise of oxygen," write Langmuir and Broecker. They even suggest that the designation of the new geological period may have to be ratcheted up to the higher level of the Anthropozoic era. An Anthropozoic era could, they warn, "be an abortive and failed mutation, as the intelligent species destroys itself and its environment." "Should we fail," write Langmuir and Broecker, "and another form of intelligent life comes along in a few tens of millions of years, they would find a planet devoid of most of its treasure chest," and a "second effort at planetary civilization would be correspondingly more difficult."[82]

Similarly, the concluding pages of a book on the deep-time dimensions of the Anthropocene written by earth scientist and paleoclimatologist Andrew Glikson and primatologist and mammologist Colin Groves contain the following warning:

> It has been lost on *Homo sapiens* that, by analogy to its own life processes which depend on the oxygen-carbon cycle mediated by the lungs, so does the biosphere depend on the planetary oxygen and carbon cycle. The phenomenon of a mammal species perpetrating a mass extinction defies explanation in terms of Darwinian evolution. . . .

Having lost a sense of reverence towards Earth, there is no evidence humans are about to rise above the realm of perceptions, dreams, myths, legends, and denial.... With a majority oblivious to the fast changing climate, disinformed by vested interests and their media outlets, betrayed by cowardly leaders and discouraged by the sheer magnitude of the event, beyond human power,... humanity is drifting into unparalleled catastrophes.[83]

And they give a name to this catastrophic process: planeticide.[84]

I can imagine many scholars in the social sciences wanting to take Glikson and Groves to task for either making *Homo sapiens* the subject of a possible planetary tragedy or for seeing the whole of humanity as "one." Some might even object to the "catastrophism" of their prose. In the hands of many social scientists, as we have seen, the subject being indicted would be different—class, developed nations, patriarchal decision structures, capitalist accumulation, European empires and colonization of lands and peoples, and so on. Some, like Christophe Bonneuil and Jean-Baptiste Fressoz, might even question the power and authority that scientists claim for themselves in defining the Anthropocene: "This then is a prophetic narrative that places the scientists of the Earth system, with their new supporters in the human sciences, at the command post of a dishevelled planet and its errant humanity. A geo-government by scientists!"[85] They oppose "handing full powers to the experts and losing the specific resources that every community has, which in their diversity and local attachments are essential motors for a just ecological transition."[86] At the other extreme, there may be those who want to see in the Anthropocene an opportunity for humans to redeem themselves by becoming effective stewards of the planet, a kind of God species.[87]

These diverse human concerns are entirely legitimate, including— especially if the scientists are not given any uncontested authority to define the problem of anthropogenic global warming—even the concerns of the so-called climate-change deniers. Faced not only with planetary environmental problems but also with enormous inequities of the human world, it is only reasonable for humans to debate their options: the pace of transition to renewable energy, geoengineering, climate-justice issues, sequestering of carbon, harvesting of rainwater, food security, climate refugee policy, adaptation and mitigation measures, and other related issues. Whether humans in the end will necessarily continue to improve and be able to prove themselves a "wise" species is a question reminiscent of a joke that Kant tells in his *The Conflict of Faculties*: "A doctor who consoled his patients from one day to the next with hopes of a speedy convalescence, pledging to one that his pulse beat better, to another an improvement in his stool, to the third the same regarding his perspiration, etc., received a visit from one of his friends. 'How is your illness, my friend,' was his first question. 'How should it be? I am dying of improvement, pure and simple!'"[88] Kant, as is well known, made his hope for human progress conditional on a number of factors: (a) "instruction [of humanity] by repeated experience," (b) "the condition of a wisdom from above" (Providence), and (c) "the prospects of an immeasurable time, provided [he said, with an eye on the history of evolution of life] at least that there does not, by some chance, occur a second epoch of natural revolution which will push aside the human race to clear the stage for other creatures like that which . . . submerged the plant and animal kingdoms before men ever existed."[89] Whether humans still have the prospects of "an immeasurable time" is, of course, a moot point in the present debate over climate change.

Bonneuil and Fressoz fear that geologists and scientists who look on global warming as both a geological event and a "human" responsibility or a responsibility of *Homo sapiens* will destroy politics. "What is left for a politics on the geological scale to which the Anthropocene summons us?" they ask. "What can we still do on the individual and collective scale given the massive scale of the Anthropocene? The risk is that the Anthropocene and its grandiose time frame *anaesthetize* politics. Scientists would then hold a monopoly position in both defining what is happening to us and in prescribing what needs to be done."[90] They feel, like Kant, that "in the face of the omnipotence of nature . . . the human being is . . . but a trifle."[91] What they overlook, however, is that their indictment of consumerism and capitalism shares the same temporal ground with arguments that look for a solution to the Anthropocene in policies advocated by climate science and a collective sense of responsibility (as at the Paris climate negotiations, for instance, in 2015). For all their differences, these different positions situate the discussion exclusively in the time of world history.

One can see at work the process of displacement that would make the time of geology obscure—"humans as a geological force" and the Anthropocene are here themes traversed by questions of power and responsibility. The displacement, first of all, substitutes for the very distributed agency (to speak with Latour again) of Earth processes some kind of a single and autonomous figure of agency (whether it is a unified figure of humanity or a particular class does not matter) to which both culpability and responsibility may be assigned. The agent here is always in a relationship of synecdoche to the distributed agency of the earth processes. In other words, the mode of being in which humans collectively may act as a geological force is not the mode of being in which humans—individually and collectively—can become conscious of being such a force. The talk of a "conscious" or responsible "force" collapses—ahead of any actual histories allowing for such a fusion—the two different modes of being human.

The displacement entailed here could be described as follows. If Earth System Science was about producing and observing planetary processes (of which intelligence would be a part) and about thus describing not a subject (human, class, and so on) but some kind of an "it" that was plural in its internal construction—the planet as an unstable system made up of imperfectly interlocking processes (including the human as a planetary force)—the place of that "it" is now taken by a subject, an "I." This is reminiscent of Lacan's analytical take, using Freud on the nature of the subject: "*Where it was*, the *Ich*—the subject . . . must come into existence."[92] Bonneuil and Fressoz describe—with reason—Earth System Science as a "a view from nowhere" (though humans now cognitively inhabit this nowhere) and ask, "What if 'Earth seen from nowhere' and the narrative of 'interactions between human species and the Earth system' were not the most interesting perspective for relating to what has happened to us in the last two and a half centuries, not to mention predicting the future? Perhaps we should accept the Anthropocene concept without succumbing to its dominant narrative[,] . . . without handing full powers to the experts." The Earth system scientists are good at "alerting us" of danger, but "they are 'from the other side,'" they say, quoting words from René Char's 1949 poem, "Les inventeurs."[93] Their words help us to see the second displacement in operation—something that needs to happen if the abyssal (for humans) time of geology were to be written over by the time of human concern. The "inside-out" perspective of human combatants of power and resistance replaces the "outside-in" point of view of Earth System Science. If we imagine Earth system scientists as—in a Latourian vein—spokespeople for the "Earth system," the act of folding back into the world-historical time of humans the geobiological time of the planet's history effects another fascinating shift. It is as if the Earth system, the planet, were saying to the conscious part of its constituents, humans—to borrow again from Lacan's language—"you never look at me from the place from which I see you."[94]

Geological Time, the Everyday, and the Question of the Political

The Anthropocene, as Nigel Clark puts it bluntly, "confronts the political with forces and events that have the capacity to undo the political." He invites humanists to "embrace the fully *inhuman*" in their thoughts, putting them "in sustained contact with times and spaces that radically exceed any conceivable human presence."[95] The Anthropocene, in one telling, is a story about humans. But it is also, in another telling, a story of which humans are only parts, even small parts, and not always in charge. How to inhabit this second Anthropocene so as to bring the geological into human modes of dwelling are questions that remain. It could indeed take "decades, even centuries," Jasanoff warns, "to accommodate to . . . a revolutionary reframing of human-nature relationships."[96]

As I have tried to demonstrate, one obstacle to contemplating such accommodation — and the related question of human vulnerability — is the attachment in much contemporary thought to a very particular construction of "the political" while the task may be, precisely, to reconfigure it. This attachment functions as a fearful and anxious injunction against thinking the geobiological lest we end up "anaesthetizing" or "paralyzing" the political itself.[97] Humans cannot afford to give up on the political (and on our demands for justice between the more powerful and the less), but we need to resituate it within the awareness of a predicament that now marks the human condition. Political thought has so far been humanocentric, holding constant the "world" outside of human concerns or treating its eruptions into the time of human history as intrusions from an "outside." This "outside" no longer exists. What is "just" for humans over one period of time may imperil our existence over another. Besides, Earth System Science has revealed how critically entangled human lives are with the geo-biochemical processes of the planet. Our concerns for justice cannot any longer be about humans alone, but we don't yet know how to extend these concerns to the universe of nonhumans (i.e., not just a few species). There is also the task of having to bring within the grasp of the affective structures of human-historical time the vast scales of the times of geobiology that these structures do not usually engage. Our evolution did not prepare us for these tasks either, as the biologist David Reznick explains.

> One useful perspective for envisioning what "sudden" means in geology is to think about how the world is changing today. We are in the midst of the sixth mass extinction. One hundred million years from now, the fossil record of our time will reveal dramatic evidence of the dispersal of humans . . . around 100,000 years ago, . . . the spread of agriculture beginning around 10,000 years ago, the advent of the industrial revolution, then the super-exponential growth of the human population. The current extinction event began during the Pleistocene with the beginning of the decline of the mammalian megafauna.

... Then there was a global decline of forests, expansion of deserts and grasslands, accumulation of industrial wastes, and an accelerating rate of extinction. ... The reason why we do not sense cataclysm, even though the geological record is certain to preserve it this way, is because of the difference in the time frame of our lives versus the time frame of the geological record. To us, 100 years is a long time. In the fossil record, 100,000 or even a million years can appear as an instant.[98]

One can see the attractions today of folding the narrative of climate change into the familiar structures of intrahuman concerns of the political that have been part of modernity since the seventeenth century and that were extended and deepened in the era that saw great waves of decolonization, civil liberties movements, feminist movements, agitations for human rights, and globalization. But all that was before the news of anthropogenic climate change broke in on the world of humanists. Anthropocene time puts pressure on another question: What does it mean to dwell, to be political, to pursue justice when we live out the everyday with the awareness that what seems "slow" in human and world-historical terms may indeed be "instantaneous" on the scale of earth history, that living in the Anthropocene means inhabiting these two presents at the same time? I cannot fully or even satisfactorily answer the question yet, but surely we cannot even begin to answer it if "the political" keeps acting as an anxious prohibition on thinking of that which leaves us feeling "out-scaled."[99]

Our sense of the planet has been profoundly based on what Edmund Husserl once famously called the "ontic certainty" of the world that human beings enjoyed. "The world is pregiven to us," he wrote, "the waking, always somehow practically interested subjects. . . . To live is always to live-in-certainty-of-the-world. Waking life is being awake to the world and of oneself as a living in the world, actually experiencing [*erleben*] and actually effecting the ontic certainty of the world."[100] He would repeat the point in his short essay on "The Origin of Geometry," the famous 1936 text that was included as appendix to his Vienna lectures of 1934.[101] The earth that corresponds to our everyday world horizon cannot be an object of any objective science.

Jacques Derrida quotes from a Husserl "fragment" titled (in English translation) "Fundamental Investigations on the Phenomenological Origin of the Spatiality of Nature," in which Husserl makes a distinction between the Copernican view of the world—(embodying some of the "planet-centered" view that Zalasiewicz mentioned) in which "we Copernicans, . . . men of modern time, . . . say the earth is not 'the

Whole of Nature,' it is but one of the planets in the indefinite space of the world"—and our everyday relationship to the earth. "The earth as a spherical body . . . certainly is not perceptible as a whole, by a single person and all at once," he remarks. It is perceptible only "in a primordial synthesis as the unity of singular experiences bound to each other" though "it may be the experiential ground for all bodies in the experiential genesis of our world-objectification." This Earth, Husserl asserts, cannot move: "It is on the Earth, toward the Earth, starting from it, but still on it that motion occurs. The Earth itself, in conformity to the original idea of it, does not move, nor is it at rest; it is in relation to the Earth that motions and rest first have sense." The unity of this primordial Earth arises out of the unity of all humanity. Even if we looked at the earth from another planet, then we would have "two pieces of a single Earth with one humanity," for, as Derrida remarks, "the unity of all humanity determines the unity of the ground [the earth] as such."[102]

Climate change challenges this ontic certainty of the earth that humans have enjoyed through the Holocene epoch and perhaps for longer. Our *everyday thoughts* have begun to be oriented—thanks again to the current dissemination of geological terms such as the Anthropocene in public culture—by the geological fact that the earth that Husserl took for granted as the stable and unshakable ground from which all human thoughts (even Copernican ones) arose actually has always been a fitful and restless entity in its long journey through the depths of geological time.[103] It is not that we have not known of catastrophes in the geological history of the planet. We have, but the knowledge did not affect our quotidian sense of an innate assurance that the earth provides a stable ground on which we project our political purposes. The Anthropocene disturbs that certainty by bringing the geological into the everyday. Nigel Clark makes this observation one of the starting points for his fascinating book *Inhuman Nature* by noticing how scientific facts can never entirely displace the "visceral trust in earth, sky, life, and water" that humans come to possess. And yet see how all four of Clark's terms are under question today: we do not know whether the earth (or Earth system) will honor our trust as we warm her up by emitting greenhouse gases into the sky; whether fresh water will run short; and whether life, as some predict, will be threatened with a sixth great extinction.[104]

Wittgenstein once said, "We see men building and demolishing houses, and are led to ask: 'How long has this house been here?' But how does one come on the idea of asking [that] about a mountain, for example?"[105] We did not ask for ages of mountains because we took them to be a part of the givenness of the earth for humans. But that sense of givenness is under challenge. Perhaps I can now provide a historian's answer to Wittgenstein's question. A time has come when the geological and planetary press in on our everyday consciousness as when we speak of there being "excess" carbon dioxide in the atmosphere—"excess" only on the scale of human concerns—or of renewable and nonrenewable sources of energy (nonrenewable on human timescales). For humanists living in such times and contemplating the Anthropocene, questions about histories of volcanoes, mountains, oceans, and plate tectonics—the history of the planet, in short—have become as routine in the life of critical thought as questions about global capital and the necessary inequities of the world that it made.

* 8 *
Toward an Anthropological Clearing

In the end, one does not know what concept one should have of a species so taken with its own superiority.

Immanuel Kant, "Idea for a Universal History with a Cosmopolitan Intent" (1784).

I began this book with the proposition that we are no longer purely in the age of the global that could be seen as the logical and historical end of modern European empires. The intense pace of globalization, extractive capitalism, and rapid evolution of technology in the decades since the Second World War have revealed to even students of the human condition the workings of the Earth system, hitherto a domain for the specialist. The Anthropocene hypothesis is about our interference with planetary processes that play a critical role in making this planet habitable for complex life. The planet offers us a perspective on humans that questions the usual assumptions that underlie the way modern humans—in their conception of themselves as modern—relate to the earth. The growing and yet sudden awareness on the part of humanists of deep, planetary history that popular-scientific literature has generated—all happening within a space of a decade—could be likened to the Heideggerian concept of the experience of "thrownness." It delivers the shock of the recognition of the otherness of the planet itself even as we regard the world-earth as *our* place of dwelling: an awakening to the consciousness that we are not always in practical and/or aesthetic relationship with the planet and yet, without it, we do not exist.

The planet destroys, as I suggested in chapter 3, the usual assumption of a relationship of mutuality between humans and the "earth," the place where they find themselves. I use the word *mutual* in its fourteenth and fifteenth century meaning of "reciprocally given and received" and not in its later seventeenth-century meaning of held or experienced in common, as may be found, for example in the expressions "mutual society" or "mutual funds."[1]

This idea of an assumed mutuality between humans and nature exists in the texts of many philosophers, though not by that name. It acquires the status of an *a priori*, for example, in Kant's *Critique of Judgment*. In his introduction to the book, Kant writes, "We have to form a connected experience from given perceptions of nature containing a maybe endless multiplicity of empirical laws, and this problem has its seat *a priori* in our understanding."[2] As the Kant scholar Douglas Burnham remarks, "nature exists [in Kant's philosophy] *as if* it were to be understood by humans."[3] Every time they look at it, said Heidegger, the earth rises up to greet humans. Human worlds and the earth are in a relationship of strife and are yet mutually bonded. To repeat him, "The opposition of world and earth is a striving."[4] The word *striving*—with its connection, in English, to the word *strife*—acts as a reminder that the relationship of mutuality between individual humans and the earth *was not necessarily one of harmony* and would include moments of "equipmental breakdown," moments characterized by what Heidegger—and Kierkegaard before him—would call anxiety or dread.

Modern Europeans discovered "deep time" in the eighteenth century—the time of geology and biological evolution—and yet modernity has been about forgetting it at the same time or as simply treating it as a background to human dwelling on the planet. Events of deep time have not so far fundamentally affected the experienced time of mutuality between the earth and humans. But this assumed relationship is under stress. What we took as the immobile—in human time—background to human action is now changing because of human action and endangering humanity. As Benjamin would have said about a moment of danger, it is as we get thrown into the abyss of deep time that an alternative history of modernity "flashes up" before our eyes and with it the possibility for a new understanding of the pasts of humans.[5]

If today humans are indeed overpowering some natural forces to such an extent that the "life-support system" of the planet—the system that supports all life, not just human life—gets increasingly broken, ushering us possibly into the beginnings of a sixth great extinction of life, what would it mean for humans—finite and individual humans—to face the planetary aspects of their own lives? We have moved on from the times when Tagore expressed his planetarity in a poetic register, or Vemula his—about a hundred years after Tagore—in a utopian-political register (chap. 5). How would humans experience their individual, finite, and singular lives today while becoming aware of the destruction of life that human activities are causing, a problem Tagore or Vemula were not called on to ponder? Here, of course, I mean life as a metaphysical and indefinable category often understood as the point at which chemistry passes into biology, something that has in deep time passed through a variety of forms, from the microbe to the megafauna. And yet this category allows us to make statements such as "life has survived five great extinctions so far" or ask whether there is life in other places in the universe.[6] The crisis registered at the metaphysical level of defining life is reflected in our everyday vital sense of life in many forms, not least in the increasingly important question of migration and refugees, both human and nonhuman.[7]

The Underpinnings of Mutuality

As we saw in the second and sixth chapters of this book, many of the great and the so-called axial religions of the world have promoted ideas about humans being in a special relationship of mutuality with the earth. But I begin with a secular and modern form of this discussion in the first half of the twentieth century. The relationship between "man" and the earth or the world became a matter of intense discussion in the years of the Great War and between the two World Wars, particularly in the late 1920s and 1930s when the question of "Man" and his civilization was raised by many modern thinkers of Europe and elsewhere. In Europe, German thinkers often took the lead on this. Max Scheler, Helmuth Plessner, Karl Jaspers, Martin Heidegger, Sigmund Freud, and others come to mind. The preeminent biographer of Sartre, Annie Cohen-Solal, documents the six years in the 1930s when Sartre immersed himself in German thought, primarily Husserl's and Jaspers's and later Heidegger's.[8] But the reach of these ideas could be felt far beyond Europe. India's Rabindranath Tagore delivered the Hibbert Lectures at Oxford University touching on these themes in May 1930. Tagore's lectures were published in 1931 as *The Religion of Man* and a few years later in a revised form in Bengali as *Manusher dharma*. There is no annotated version of Tagore's lectures, but he must have been influenced by contemporary thought.[9] Some similarities—allowing for deep differences in training—between his ideas and those of European thinkers pursuing questions of philosophical anthropology in the years between the two devastating world wars of the twentieth century are indeed striking.[10]

I will begin my discussion here with this text of Tagore's. Tagore was not, of course, a trained philosopher or theologian or a systematic thinker. He said in his own defense, "my religion is a poet's religion. All I feel about it is from vision and not from knowledge. Frankly,

I acknowledge that I cannot satisfactorily answer any questions about evil, or about what happens after death."[11] Moreover, the politics of his lecture were actually about critiquing aggressive forms of imperialist and nationalist ideologies. But some critical parts of his lectures help us identify at least three salient strands and assumptions broadly common in the thoughts of that time. They are also the principles that underlie what I call here the structure of mutuality.[12] These assumptions are (a) the specialness of humans, (b) the centrality of humans to the larger scheme of things, and (c) the idea that humans have the capacity to have visions of the whole world as totalities, albeit of different kinds. In the discussion below, I will retain the capitalized words *Man* and *Nature* that Tagore and his contemporaries used to refer to the whole of humanity and nature, respectively.

THE SPECIALNESS OF MAN, OR MAN AS EXCEPTION

Tagore leveraged his reading of the history of the biological evolution of *Homo sapiens* to make some powerful and secular claims as to why of all creatures, humans were the most special. The relevant part of his lecture began with what we would today call "Big History:" "Light, as the radiant energy of creation, started the ring-dance of atoms in a diminutive sky.... The planets came out of their bath of fire and basked in the sun for ages.... Then came the time when life was brought into this arena." But in the history of life, Man was special: "Before the chapter [of evolution] ended Man appeared and turned the course of this evolution from an indefinite march of physical aggrandizement to a freedom of a more subtle perfection."[13] While "the development of intelligence and physical power" was "equally necessary in animals and men for their purposes of living," what was "unique in man" was "the development of his consciousness which gradually deepens and widens the realization of his immortal being."[14]

Tagore's argument based itself on the history of bipedalism as he knew it. His knowledge led him to see bipedal existence as exclusive to humans. The evolutionary history he rehearsed was wrong. Bipedalism was a trait of hominins ("by 5 million years ago at the latest," before the evolution of genus *Homo* about 1.8 million years ago), and there were even prehistoric apes who showed a degree of this development.[15] But this is how Tagore argued, anyway: "In the very beginning of his career Man asserted in his bodily structure his first proclamation of freedom against the established rule of Nature. At a certain bend in the path of evolution he refused to remain a four-footed creature, and the position which he made his body to assume carried with it a permanent gesture of insubordination," said Tagore.[16] He was aware that being bipedal was not "natural" for an ape: "For there could be no question that it was Nature's own plan to provide all land-walking mammals with two pairs of legs, evenly distributed along their lengthy trunk heavily weighted with a head at the end. This was the amicable compromise made with the earth when threatened by its conservative downward force, which extorts taxes for all movements." But Man defied Nature's edict as a matter of spiritual self-assertion. Tagore continued: "The fact that Man gave up such an obviously sensible arrangement proves his inborn mania for repeated reforms of constitution, for pelting amendments at every resolution proposed by Providence."[17]

Tagore's arguments were not made without humor. "If we found a four-legged table," he wrote teasingly, "stalking about upright upon two of its stumps, the remaining two foolishly dangling by its sides, we should be afraid that it was either a nightmare or some supernormal caprice of that piece of furniture, indulging in a practical joke upon the carpenter's idea of fitness." He acknowledged the similar "absurdity" of the human ape's preference for standing upright on its two feet: "The like absurd behaviour of Man's anatomy encourages us to guess that he was born under the influence of some comet of contradiction that forces its eccentric path against orbits regulated by Nature." Tagore also knew something of the physical price that humans paid for their adoption of bipedalism in their evolutionary history. He remarked, "And it is significant that Man should persist in his foolhardiness, in spite of the penalty he pays for opposing the orthodox rule of animal locomotion. He reduces by half the help of an easy balance of his muscles. He is ready to pass his infancy tottering through perilous experiments . . . upon insufficient support, followed all through his life by liability to sudden downfalls resulting in tragic or ludicrous consequences from which law-abiding quadrupeds are free."[18] But this was a price that Tagore's Man was prepared to pay for his freedom.

MAN AT THE CENTER OF THINGS

Tagore put it simply: "The capacity to stand erect has given our body its freedom of posture, making it easy for us to turn on all sides and realize ourselves at the centre of things."[19] The world exists as if mainly for us. "Somewhere in the arrangement of this world there seems to be a great concern about giving *us* delight, which shows that in the universe, over and above the meaning of matter and forces, there is a message conveyed through the magic touch of personality."[20] "When I was eighteen," he wrote when he was about seventy, referring to a moment of epiphany in his youth, "a sudden spring breeze of religious experience for the first time came to my life. . . . One day while I stood watching at early dawn the sun sending out its rays from behind the trees, I suddenly felt as if . . . the morning light on the face of the world revealed an inner radiance of joy."[21]

MAN AS A VIEWER OF THE WHOLE

The third assumption, that the world presented itself as a whole only to Man, simply followed from what Tagore had said above. He clarified, though, that this was not the natural "world" that scientists worked on: "It is not the world which vanishes into abstract symbols behind its own testimony to Science, but that which lavishly displays its wealth of reality to our personal self[,] having its own perpetual reaction upon our human nature."[22]

To drive his point home, Tagore focused on the function of human eyes: "from the higher vantage point of our physical watch-tower we have gained our *view*, which is not merely information about the location of things [animals also get that] but their inter-relation and their unity."[23] This passage in the printed text was a developed version of what Tagore had said in the first lecture that he actually gave to his audience at Oxford: "The awareness of objects which animals obtain through smell and sight is essentially in the interest of immediate needs. By lifting up his head, man no longer saw merely separate and distinct objects; he also had a complete view of the unity of manifold things. He saw himself at the centre of an undivided extention. The erect prized the distant more than the near."[24] The closely related nature of the second and the third assumptions about Man would be obvious from these quotations.

The Materially Empty World of Mutuality

Ironically but typically, the "world" that Tagore saw "Man" as being in a deep relationship with was very lightly loaded with materiality. The relatively empty but capacious category of "the world"—in its relationship to the human—came to absorb and erase into its vast and vacuous oneness the rich, strange, and intractable diversity of what actually exists. (Tagore's poetry or songs, on the other hand, would not be open to this criticism.) This is an experience of the world bereft of any sense of the alterity of the planet that could have arisen from the historical sense of deep time; what remained was a metareligious sense of an unchanging infinity in which structures simply repeated themselves endlessly. Here is Tagore again, writing around 1914/1915: "The most surprising thing of all [is] . . . how ceaseless is the fountain of forms that issues constantly from the formless One. . . . I have observed that the sun shines brighter and the light of the moon feels heavier with sweetness when my heart feels full of love.... From this I know, the world and my mind and heart are inseparable."²⁵

Tagore's own version of mutuality between humans and the world—"this world we perceive through our senses and mind and life's experience is profoundly one with ourselves"—may have been too idealistic, but the structure of assumptions we discussed above may be seen at work in many humanist texts as themes that have endured across significant divides of space and time.²⁶ They become particularly marked in aestheticized and spiritualized experience of landscapes, vulgar and vulgate versions of which often turn up in glossy guidebooks for world travelers, books that freeze the tumultuous movements of geological history into the human-aesthetic category of the landscape. Indeed, one could argue that the aesthetic-spiritual category "landscape" acts as a screening device in the mechanism of what I have called "mutuality": it hides from us the catastrophic richness of the contingent histories of geology and life.

Take, for instance, a passage from Martin Hägglund's recent book, *This Life: Secular Faith and Spiritual Freedom*, a book that stridently argues against religion. "One late summer afternoon," writes Hägglund, "I am sitting on top of a mountain in northern Sweden. The ocean below me is calm and stretches towards an open horizon. There is no other human being in sight and barely a sound can be heard. Only a seagull is gliding on the wind." Hägglund has been here before, but every encounter is fresh and unique. The world is a structure that both repeats and constantly refreshes itself: *"Like so many times before,* I find it mesmerizing to follow a seagull as it hovers in the air and lingers over the landscape. For as long as I can remember, seagulls have been a part of my life.... Yet *I have never encountered a seagull the way it happens this afternoon.* As the seagull stretches its wings and turns toward an adjacent mountain, I try to imagine how the wind may feel and how the landscape may appear *for the seagull."*²⁷ Of course, Hägglund quickly acknowledges that he will "never know what it is like to be a seagull." Yet in his mind, this solitary encounter with another creature—and the vastness of a seemingly empty landscape is what makes it "solitary" in this case—gives rise, in a Heideggerian fashion, to a fundamental question of being: he wonders "what it means to be a seagull."²⁸

As the reader will notice from Hägglund's passage, his sense of a relation of mutuality with the earth refers to the structure of an experience that can repeat itself: "like so many times before." Mutuality arises for a single human being who faces what surrounds her or him from within the solitariness of her or his singular human life and who experiences the surroundings not only as rising up to meet her gaze but also—and equally importantly—as stable. For it is the stability of the landscape that allows for the experience to be repeated. William James assumed this idea of mutuality when he said quite early in his series of lectures on varieties of religious experience, "Religion . . . shall mean for us the *feelings, acts and experiences of individual men in their solitude, so far as they apprehend themselves to stand in relation to whatever they may consider divine.*"[29] Martin Buber's classic and celebrated conception of the I-Thou relationship is yet another instance of what I am calling mutuality within which the bearer of singular, finite human life exists.[30]

In aesthetic-spiritual experiences of landscapes, it is the landscape that the human stands "in relation to," and it takes the place of the divine in James's schema. Charles Taylor's idea of "fullness of human life" that he sees as critical to this "secular age" has as its point of reference the experience of singular lives, as otherwise the question of "experience" could not arise. Taylor quotes from the autobiography of the British-born Benedictine monk and yogi Bede Griffiths (*The Golden String* [1979]) a passage that bears remarkable similarity to the passage I have quoted from Hägglund, with about four decades separating the one from the other. Once, "walking out alone in the evening" while he was still in school, Griffiths experienced the song of birds in a strikingly new manner:

> A lark arose suddenly from the ground beside the tree where I was standing and poured out its song above my head, and then sank still singing to rest. Everything that grew still as the sunset faded and the veil of dusk began to cover the earth. . . . A feeling of awe . . . came over me. I felt inclined to kneel on the ground, as though I had been standing in the presence of an angel; and I hardly dared to look on the face of the sky, because it seemed as though it was but a veil before the face of God.[31]

Griffiths's reference to "faces"—of God, of the sky—speak of that reciprocal act of facing that comes to be seen as in a relation of contiguity to the divine.

We could move even farther back to a journal entry of Søren Kierkegaard of July 29, 1835, where he wrote, "When walking from the inn over Sortebro [Black Bridge] (so called because the bubonic plague was supposedly checked there) to the open ground along the beach, almost a mile north one comes to the highest point around here. . . . *This has always been one of my favorite spots.*" "*Often*, as I stood here of a quiet evening, the sea intoning its song with deep but calm solemnity, my eye catching not a single sail in the vast surface, and only the sea framed the sky and the sky the sea, and when, too, the busy hum of life grew silent and the birds sang their vespers." Notice the structure of repetition evoked by the word *often* and the word *frame* referring us back to paintings of landscapes. Then the process of what I have described before as the thinning or emptying of the "world" takes over. Kierkegaard would not have been alone in that landscape. There were other creatures around him—he himself acknowledges the presence of some birds—and yet they fall out of his view as darkness descends as if to uphold an invisible structure of mutuality between this man and the world he faced: "As I stood there, . . . alone and forsaken, the power of the sea and the battle of the elements reminded me of my nothingness, while the sure flight of the birds reminded me on the other hand of Christ's words, 'Not a sparrow will fall to the earth without your heavenly Father's will[.]' I felt at once how great and yet how insignificant I am."[32]

* * *

The problem of mutuality, of course, does not have to entail the question of facing in a simplistic manner. One could understand the problem in a Levinasian way, not as one face turned necessarily toward another but as the face itself being constituted through an exposure to alterity.[33] The more fundamental point is that the structure of mutuality can only work when the word *life* refers to the singular life of the finite individual, for the ontological projection within which the "experiencing" of mutuality with the earth can actually happen is a question that arises only through the case of individual humans. Once expressed and shared, however, these experiences can collectively form a community of sentiments and experiences. What happens, however, to this sense of mutuality when we are summoned by the general crisis of life—the possibility of another great extinction—to the impossible task of witnessing the convulsive geological history of the planet in all that is critical to the flourishing of our own and other forms of life?

Seeing the Planet in the Earth

The planetary involves the work of deep time. We do not normally think of deep time. It remains as a part of what we take to be given. Recall Wittgenstein's question from chapter 7: We ask of buildings how old they are, why do we not ask the same of a mountain? We do not do so, it seems, because we think of the mountain—or the landscape—as simply providing the background against which to experience our relationship of mutuality with the earth. The rocks are a part of the landscape. What if we could see the landscape itself as being in movement—often cataclysmic—in deep and historical time: seas rising to submerge land or droughts ravaging them, extractive capitalism producing "dead zones" in the seas and on land, with species habitat getting destroyed, with landscapes no longer constituting merely a background for human action?[34] The noted photographer Edward Burtynsky has captured a striking image of the relationship between humans and landscape. A couple enjoys the sun on the rock-cradled Itzurun Beach in Spain's Basque Country. Behind them, unheeded, the layers of rock bear witness both to the mass extinction sixty-five million years ago that destroyed the dinosaurs and to the major warming event (PETM) that happened ten million years later and lasted two hundred thousand years. When will the couple notice the past, preserved in rocks around them?[35]

Here is a problem of the kind that J. B. Haldane discussed in his famous 1926 essay, "On Being the Right Size."[36] Humans' lives, whether aided by modern medicine and public health or not, usually span a few decades. That is the phenomenological ground on which we stand. Even a few thousand years, not to speak of millions, are just too vast for our experience. This is what allows for the structure of repetition that marks the idea of mutuality with the earth, as illustrated by the theme of returning to the same landscape in one individual life that surfaced in what we cited from Kierkegaard, Griffiths, and Hägglund above. Imagine an impossible scenario: suppose individual humans could live on geological or unhuman scales of time. What would have happened to mutuality if we all, as individuals, lived for twenty thousand years? The structure of repetition that I discussed while commenting on the quotations from Hägglund, Griffith, and Kierkegaard would have been impossible to achieve. They could not have spoken of nature spots to which they returned every so often to experience the same experience! If they had each lived for more than twelve thousand years—not a large number on geological scales of time—they would have seen how unstable the landscape was, changing its form every so often because of cataclysmic geomorphological changes. There would not have been any stable features to return to. The structure of mutuality becomes possible because human time, being what it is, allows us to be forgetful about deep time by unconsciously converting it into a figure of space: the seemingly enduring landscape.

But this is what is changing. It is as if the crisis of the Anthropocene—the prospect of inhabiting a less habitable planet—reduces us to our creatureliness, the state of "the first men," as often imagined by Enlightenment philosophers of human civilization, such as in the "seventh and last epoch" in Buffon's history of the making of the world: "The first men, witnesses of convulsive movements of the Earth, then still recent and very frequent, having only mountains as refuge against inundations, often chased from these same refuges by the fires of volcanoes, trembling on an Earth that was trembling under their feet, exposed to the curses of all elements, ... all equally penetrated by a common feeling of baleful terror."[37] Today, the work of deep time is beginning to break into our everyday consciousness of human-historical time, calling on us to witness, like Buffon's first men, the convulsive nature of this planet. That is *the* "shock of the Anthropocene," signaling a breach in the structure of human-historical time and in the structure of mutuality and compelling an engagement with deep time and the history of life on this planet. Eugene Thacker puts his finger on the problem, though it arises for him in a related but different context:

> If the existence of disasters, pandemics, and nonhuman networks tells us anything, it is that there is another world in addition to the world that is there "for us." This is not simply a world in itself, and neither is it a world that is destined for us—rather, it is a world that presents us with the very limits of our ability to comprehend it in terms that are neither simply that of the "in itself" or of the "for us." It is a world "without us" (the life *sans soi*). It is the challenge of thinking a concept of life that is foundationally, and not incidentally, a nonhuman or unhuman concept of life.[38]

Following Thacker, I could ask, What would be our ethical-spiritual relationship to the planet—neither an "in itself" nor "for us"—that refuses to grant us the usual assurance of an imagined relationship of mutuality with the earth that, while itself older than modernity, has also accompanied our sense of being modern?

Relating to the Planet That Is the Earth System

For an answer to the question above, I turn to planetary scientists themselves, for clues here are provided—just as we saw in chapter 3—by the very deeply phenomenological and human responses that their cognitive encounter with the planet or the Earth system produces in them. Many of these responses, for instance, gather around the question of geoengineering, the various plans that are now being seriously considered for managing the climate of the whole planet if not the Earth system itself.

Those who champion geoengineering belong as a rule to sciences that are ahistorical in their analytical approach—such as physics and chemistry. Those who study the planet historically, such as geologists or evolutionary biologists, are usually wary of such measures.[39] I should be absolutely clear that I am not weighing in on the ongoing debate about the desirability or otherwise of geoengineering. I am not qualified to do that. I am simply reporting on the very different spiritual relationship to the category planet (or Earth system) that these sciences seem to prompt in their individual practitioners. In my terms, it is as if, as individual humans, some of these scientists struggle to fold the planetary into the global and thus into a structure of mutuality.

The Harvard physicist David Keith's lucid and engaging defense of "climate engineering" in his popular book on the subject provides one example.[40] One of his founding premises is that humans care about nature (mutuality) and that geoengineering is a continuation, precisely, of that care. "A fuzzy love of nature," writes Keith, "is uncontroversial." "I suspect," he adds, "that Edward O. Wilson, the entomologist and writer, captured more than a grain of truth with his biophilia hypothesis, that humans have an innate urge to affiliate with other forms of life." There is thus, for Keith, no conflict between caring for nature and the project of geoengineering so long as one does not make "naïve claims of a sharp distinction between nature and civilization."[41] There is no Nature untouched by human activity, so climate or geoengineering is simply a matter of humans being able to manage an expanded version of that reality.

Keith's view of geoengineering as an expression of "biophilia" is significant. Biophilia belongs to the structure of mutuality. Wilson coined the term *biophilia* in 1979 while writing an article on conservation for the *New York Times* in 1979. "It means," he wrote later after having written a book on the subject in 1984, "the inborn affinity that human beings have for other forms of life, an affinity evoked . . . by pleasure, or a sense of security, or awe, or even fascination blended with revulsion. One basic manifestation of . . . biophilia is a preference for certain natural environments as places for habitation." The idea was based on the work of Gordon Orians, a zoologist at the University of Washington who had asked people about their "ideal" habitats to discover that, given a choice, most of them wished "their home to perch atop a prominence, placed close to a lake, ocean, or other body of water, and surrounded by a parklike terrain." "The trees they most want to see from their homes have spreading crowns, with numerous branches projecting from the trunk and horizontal with the ground, and furnished profusely with small and finely divided leaves." Wilson realized that "this archetype fits a tropical savanna of the kind prevailing in Africa, where humanity evolved for several million years."[42] Biophilia, however, reproduces only the limited "world" of mutuality. It can extend only to forms of life visible to humans, not to the microbes and bacteria that make up the bulk of life. As Cary Wolfe cogently argued, it is not possible from a human point of view to give all forms of life equal value: "Will we allow anthrax or cholera microbes to attain self-realization in wiping out sheep herds or human kindergartens? Will we continue to deny salmonella or botulism micro-organisms their equal rights when we process the dead carcasses of animals and plants that we eat?"[43]

In contrast to such physicists as David Keith, the geologist Andrew Glikson recommends an attitude of reverence to the earth, though he expresses this as part of a narrative of loss. I quoted him in the previous chapter, but his words would bear repetition in the context of the current discussion. Glikson writes, "Having *lost* a sense of *reverence* toward Earth, there is no evidence humans are about to rise above the realm of perceptions, dreams, myths, legends and denial.... Perhaps it is too much to expect any living species to possess the wisdom and responsibility required to control its own inventions.... [But] without ethics, *Homo sapiens* cannot survive."[44] The geologist Marcia Bjornerud also recommends a cautionary approach to the planet. Speaking of "the idea of cooling the planet by shooting sulfate aerosols in the atmosphere," a measure David Keith advocates, she points out that "that sky would always be white," never blue. Imagine an earth without a blue sky and what happens to our sense of the mutuality! "The most vocal advocates for stratospheric sulfate injection," she adds, "are either economists, accustomed to viewing the natural world as a system of commodities ... or physicists, who treat it as an easily understood laboratory model.... Most geoscientists, knowing the long and complex story of the atmosphere, biosphere, and the climate ... think the idea humans can manage the planet is delusional and *dangerous*."[45] The climate scientist Wallace Broecker's words, quoted in chapter 2, may be recalled: "Every now and then ... nature has decided to give a good swift kick to the climate beast. And the beast has responded, as beasts will—violently and a little unpredictably."[46] In his recent book *Half-Earth*, Wilson himself used the word *dangerous* in referring to the "gargantuan and dangerous programs of geoengineering now being discussed" and instead proposed that to save biodiversity, humans should follow "the precautionary principle" and leave half of the land surface on the planet to forms of life other than human.[47]

Modernity and the Loss of Reverence

From where and how does a humanist historian begin to think in order to contribute to the new-political work of composing "the common" that Bruno Latour and others have proposed as the way forward without in the process denying all that already divides humans in the space of the political? Clearly, however one thinks of human futures, one condition set by European political thinkers of modernity will have to hold in any definition of the political: humans will need protection from predators. Human dwelling has always been about feeling safe. While a mere shelter, in Heidegger's philosophy, does not amount to dwelling, dwelling incorporates the principle of "shelter"; it safeguards humans:

> In what does the nature of dwelling consist? Let us listen once more to what language says to us. The Old Saxon *wuon*, the Gothic *Wunian*, like the old word *bauen*, mean to stay in a place. But the Gothic *wunian* says more distinctly how the remaining is experienced. *Wunian* means to be at peace, to be brought to peace, to remain in peace. The word for peace, *Friede*, means the free, *das Frye*, and *fry* means: preserved from harm and danger, preserved from something, to be safeguarded.[48]

This fundamental requirement of dwelling, that it should involve the question of feeling safe, goes far, far back into the deep history of humans. Writing about the South African australopithecine (a bipedal primate with both ape and human characteristics) sites, Robin Dunbar remarks that many of these were limestone caves "edging river valleys," providing these creatures both warmth and "safety from predators in areas where large trees in which to roost [were] in limited supply."[49] The need to be safe from predators could have only increased with the domestication of certain animals, the rise of agriculture, and eventually of creation of cities so that any "modern" human settlement or civilization—as may be illustrated by the history of the large-scale slaughter of dingoes by European settlers in Australia—would come to define itself as a human-dominant order of life, that is, an order defined by its capacity to keep humans safe from predators, big and small.[50]

But today it is a question of feeling safe on a planet where many areas may become uninhabitable—not just for humans but for many other species as well. Protection therefore has to be extended not simply to the citizens of a nation-state but to immigrants, refugees, and aliens whose numbers will most likely swell both within and across nations— exactly the opposite of the main trends in the anti-immigrant politics of so many nations today. And the politics of human well-being has to be in conversation with the problem of "habitability" of this planet, the awareness that human history is but a part of the history of complex life on this planet, that biodiversity is critical to making this planet habitable. We have to begin to think of humanity as not only a planet-wide diaspora of a biological species but also of this diaspora as constituting a minority form of life, the mainstay of biological life on the planet being microbial. We will have to make our way toward an order that presently seems unimaginable: an order that is not necessarily human dominant. Our new-political thought also then has to draw on the intellectual legacy of minority thinking and not issue from the position of those who assume their dominance in the order of things.[51] There are no paths here already charted out for us. We increasingly see how hopelessly humanocentric all our political and economic institutions still are. The political eventually will have to be refounded on a new philosophical understanding of the human condition.

Toward an Anthropological Clearing

I end therefore with a particular exercise in thought. It is somewhat similar to what the German philosopher Karl Jaspers undertook in the period between the two world wars of the twentieth century when he developed his idea of an "epochal consciousness." I have disagreements with Jaspers that need not detain us here. But I find in his thoughts some of the spirit of this exercise foreshadowed.

The context for Jaspers's thoughts was, of course, very different. "Nuclear holocaust" does not betoken the same kind of crisis as global warming. While both may be anthropogenic in nature, a nuclear crisis could be a one-off event of catastrophic proportions, but global warming names a series of events unfolding over the lives of many generations and beyond. Yet because of their capacity to destroy civilization as we know it, both call to our attention the problem of the common(s), and this is where Jaspers's thoughts may still have something to offer. Two aspects of the idea appear to have special relevance to what I am seeking to do here: (a) his thinking on "epochal consciousness" comes out of a particular tradition (mainly German) of taking the whole of humanity as the object of philosophy of history at a moment of global crisis or emergency, and much of my thinking here is an heir to that tradition; and (b) the fact that Jaspers invented this category "epochal consciousness" to find a home for thought that did not foreclose the space of actual politics and yet created a perspectival and ethical vantage point, something that he qualified as "prepolitical." Prepolitical in a very special sense: a form of consciousness that does not deny, decry, or denounce the divisions of political life while seeking to position itself as something that comes *before* politics or *before* thinking politically, as a pre-position as it were to the political.

In his book, *Man in the Modern Age*, published in German in 1931 and in English in 1933, Jaspers spelled out the idea of "epochal consciousness" as a problem that had haunted European intellectuals "for more than a century." Furthermore, he argued that it was a problem that had become urgent "since the [Great] war," from which time "the gravity of the peril [to humanity] ha[d] become manifest to everyone."[52] Jaspers explained the context for "epochal consciousness" as follows: "Man not only exists but knows that he exists. In full awareness he studies his world and changes it to suit his purposes. He has learned how to interfere with 'natural causation.'"[53] Epochal consciousness was thus a "modern" phenomenon, a phenomenon possible only after Man had learned to "interfere with 'natural causation.'"

While there had been "transcendental" and universal conceptions of history before—such as the Christian, Judaic, or Islamic ones—passed on "from one generation to another," the continuity of this chain, argued Jaspers, was "severed" in the sixteenth century with "the deliberate secularization of human life."[54] This was the beginning of the process of European domination of the globe: "It was an age of discovery. The world became known in all its seas and lands; the new astronomy was born; modern science began; the great era of technique was dawning; the State administration was being nationalized."[55] The French Revolution was perhaps the first event that found expression in forms of "epochal consciousness" in the work of philosophers. It was "the first revolution whose motive force was a determination to reconstruct life upon rational principles after all that reason perceived to be the weeds of human society had been ruthlessly picked up and cast into the flames." Even though the "resolve to set men free developed into the Terror which destroyed liberty," the fact of the revolution, wrote Jaspers, left men "uneasy about the foundations of an existence for which they thenceforward held themselves responsible, since [existence] could be purposively modified, and remoulded nearer to the heart's desire."[56] Jaspers mentions Kant, Hegel, Kierkegaard, Goethe, Tocqueville, Stendhal, Niehbuhr, Tallyrand, Marx, and, among others, Nietzsche as bearers of different forms of epochal consciousness, ending his series with Walther Rathenau's *Zur Kritik der Zeit* (1912) and Spengler's *Untergang des Abendlandes* (1918) as two books displaying forms of epochal consciousness that preceded his own on *Man in the Modern Age*.[57] And we can, of course, add to this list other names of the twentieth century including those of Martin Heidegger, Hannah Arendt, and, as the preceding discussion suggests, Rabindranath Tagore.

Epochal consciousness as a form of thought was supposed to have two characteristics. It was nonspecialist thought and, more importantly, it was not oriented to finding solutions.[58] Epochal consciousness, Jaspers said, is "granted to man without giving him the rest of a conclusion." To inhabit such consciousness "takes stamina," for "it calls for endurance in the tensions of insolubility."[59] Epochal consciousness is ultimately ethical. It is about how we comport ourselves with regard to the world under contemplation in a moment of global—and now planetary—crisis. It is what sustains our horizons of action. So I offer the following in a spirit of dialogue with the reader. As Jaspers wrote, citing Nietzsche, "Truth begins when there are two."⁶⁰

Wonder and Reverence

Recall the words the geologist Andrew Glikson: "Having *lost a sense of reverence* toward Earth, there is no evidence humans are about to rise above the realm of perceptions, dreams, myths, legends and denial. . . . Perhaps it is too much to expect any living species to possess the wisdom and responsibility required to control its own inventions. . . . [But] without ethics, *Homo sapiens* cannot survive."⁶¹ Reverence, wisdom, responsibility, ethics—these are secular words denoting a spiritual relationship to earth. There is no doubt that we are way beyond the language of geology here and are in the neighborhood of theology. Reverence is not simply about curiosity, wonderment, or biophilia. Reverence suggests a relationship of respect mixed in with fear and awe, with proto-Italic roots that mean "to be wary."⁶² We do not fully understand the planet and its processes. It does not belong to the structure of mutuality that Heidegger, Tagore, and others outlined. We cannot even always predict its "anger," so we need to be wary of it. The planet can, as demonstrated by the Australian fires of 2019, reduce us to our creaturely lives where we compete with other species (as with camels in Australia over water) for sheer survival. Watching out for something that is both miraculous (because it bears complex life) and dangerous—not always to be embraced in mutuality—this is the spirit of which Glikson speaks, though in a nostalgic and cautionary register. Nostalgic because humans appear to have forgotten that the planet has something about it of what Rudolf Otto in *The Idea of the Holy* described as "mysterium tremendum"—something "urgent, active, compelling, and alive." Capable of expressing a "wrath" that has nothing moral about it—it does not punish humans for anything they may have done wrong—it can reduce us to the sense of an abject creaturehood, so overpowering can be its presence.⁶³

Biodiversity evokes in human students of the phenomenon a sense of the miraculous, for there was nothing inevitable about the coming of complex life on this planet. The Earth system produced a delicate atmosphere—our "modern" atmosphere—allowing complex animal and plant life to flourish. That atmosphere has persisted for about four hundred million years. We depend on it, but *it was not made with us in view*. Theoretically, it would be there even if humans had not appeared. When humans came along as an animal, our animal life was full of fear—one source, one might say, mimicking William James, of a variety of ancient religious experiences. Most importantly, there was the fear of other animals and spirits, reverence toward the nonhuman and the nonliving. The ancient religions of Africa, the Americas, and Australasia do not suggest the idea that the world was created mainly or even only for humans. Fear was critical to survival of species, for fear regulated interspecies relations. Then came the Holocene and human civilization. The axial religions made us feel central to the miracle of creation. The seventeenth and eighteenth centuries—a time when Europeans seized other peoples' lands and increasingly found themselves in the lap of luxury—sometimes made European thinkers overconfident about the place and prospects of humanity. In his *Second Treatise*, Locke announced—with the New World very much in mind—land for humans to be as abundant as water: "Nor was this *appropriation* of any parcel of *land*, by improving it, any prejudice to any other man, since there was still enough, and as good left; and more than the yet unprovided could use. . . . Nobody could think of himself injured by the drinking of another man, though he took a good draught, who had a whole river of the same water left to him to quench his thirst: and the case of land and water, where there is enough of both, is perfectly the same."[64] Grotius in his *Mare Liberum* (The free sea) (1609) pronounced the oceans to be not only common property of all humans but also inexhaustible in the amount of food they contained for us: "For it is generally agreed that, if a great many persons hunt or fish upon some wooded tract of land or in some stream, the wood or stream would probably be [exhausted of] wild animals or fish, an objection which is not applicable to the sea."[65] It was 1735 when a European savant, Carl von Linné even cataloged us, his own species, under Wise Man, *Homo sapiens*, with a little note on the side, as if addressed to himself, saying: *nosce ti ipsum* (know thyself). He included it in the tenth revision of his *Systema naturae* (1758).[66] And later in the same century, Kant proclaimed with confidence that the fleece on the sheep's back was meant for us, humans.[67] Finally came industrial and capitalist modernity followed by the intense globalization and democratization of consumption of the last four decades. We gradually forgot the culture of reverence on which all ancient, Indigenous, and even peasant religions were based.[68]

Glikson is right to speak of reverence for the planet in a nostalgic tone, for becoming modern—whether in Europe or in its colonies—was fundamentally about overcoming fear in many different senses (including the fear of foreign or domestic oppressors). Readers will recall that Horkheimer and Adorno opened their *Dialectic of Enlightenment* with the following observation: "In the most general sense of progressive thought, the Enlightenment has always aimed at liberating men from fear and establishing their sovereignty."[69] Through nineteenth-

and twentieth-century waves of modernization, given the combination of electricity and technology and the rise in the number of cities and their inhabitants, humans overcame their fear of—and reverence for—other forms of life and of what they took to be a part of the givenness of their world.

To be modern, then, was to retain an Aristotelian sense of wonderment and curiosity about the world and the universe but lose all sense of fear as a value (as distinct from fear as an instinct or drive)—except for the citizen's political fear of the law and the state, as Hobbes would say. If this is true, then a task faces the historian of the Anthropocene. True, modern humans, as both a condition for and a consequence of their flourishing, have lost the fear of other species. Our life in what we call human civilization depends crucially on not having to be fearful of most other life-forms as we go about our everyday business. This is a basic condition of modern life. But how did we come to lose this fear? I ask because this did not happen everywhere at the same time. I grew up in a Calcutta of the 1950s where the fear of wild animals—mainly foxes, snakes, bats, and frogs of all varieties—was as real as that of ghosts and spirits. Later, as electricity came and people moved in, clearing land to build more houses, these creatures disappeared. The variety of birds to be seen in the city dwindled. The process is still ongoing in India. As the work of Annu Jalais among the villagers of the delta area of the Sundarbans in southern Bengal shows, there are still areas in India and Bangladesh where animals (in this case, the tiger) are still both feared and sometimes worshipped.[70]

So here is indeed a project for provincializing Europe. How does one write the history of modern humans' loss of fear—not as an instinct but as a value, a process that no doubt begins in Europe? Seventeenth-century founders of modern (European) political thought assumed that the human fear of wild animals was by definition a part of the "natural" order. Protection of human life and property in their texts, therefore, meant protection from predatory practices of other humans, not wild animals.[71] Hobbes's discussion of "rights over non-rational animals" in his *De cive* (1642) assumed, for example, that human habitations were already free of the fear of wild animals. This was a condition for being political: "In the natural state, because of the war of all against all, any one may legitimately subdue or kill Men, whenever that seems to be to his advantage; much more will this be the case against animals. That is, one may at discretion reduce to one's service any animals that can be tamed or made useful, and wage continual war against the rest as harmful, and hunt them down and kill them. This *Dominion* over animals has its origin in the *right of nature*, not in *Divine positive right*."[72] He adds, "the condition of mankind [in the state of nature] would surely have been very hard, since the beasts could devour them in all innocence, while they could not devour the beasts. Since, therefore it is by natural right the animal kills a man, it will be by the same right that a man slaughters an animal."[73] For Hobbes, then, "wild animals" were already a part of the original state of humans, not of their condition when they had built the state.

The planet now reminds us that while valuable, this pursuit of overcoming the feeling of reverence for the world around us has also been a loss in the sphere of values. And a critical loss in some ways. In building a new tradition of political thought that is not simply about human domination of the earth, we would need to find ways of combining elements of both wonderment and reverence in our relationship to the places we inhabit. While that task has to be accomplished collectively and historically and through the existing space of the political with no guarantees of success, Hobbes gives us at least a rhetorical starting point. But we find that point of departure only by reading him against his own intention. In *Leviathan*, Hobbes glosses for his readers two Latin words, *prudence* and *sapience*. He writes, "As much experience, is *Prudence*; so, is much Science, *Sapience*. For though we usually have one name of Wisedom for them both; yet the Latines did always distinguish between *Prudentia* and *Sapientia*; ascribing the former to Experience, the latter to science." Hobbes valued "science" over experience. Prudence or experience is "useful," he says, but science "infallible." He was clear: "*Reason is the pace* [peace]: Encrease of *Science*, the *way*; and the Benefit of mankind, the *end*."[74] In the absence of science, there could be the guidance of experience and "natural judgment," for the worst offense in Hobbes's eyes was sticking to texts dogmatically, the offense of pedantry: "But in any businesse, wherof a man has not infallible Science but to proceed by; to forsake his own natural judgement, and be guided by generall sentences read in Authors, and subject to many exceptions, is a signe of folly, and generally scorned by the name of Pedantry."[75] But experience was inferior to science. It did not carry the "infallibility" of science, and worse, it was no guarantee against the ambiguities of language, and for that reason it provided no basis for order. Hobbes therefore always preferred science, *Sapientia*, as something of higher value in his philosophy of political order than sheer *Prudence* or experience. "The Light of humane minds," he writes in *Leviathan*, "is Perspicuous Words, but by exact definitions first snuffed, and [then] purged from ambiguity; And on the contrary, Metaphors, and senseless ambiguous words, are like *ignes fatui* [fools' fire]; and reasoning upon them is wondering among innumerable absurdities; and their end, contention, and sedition, or contempt."[76] He defines *Prudence* as "a *Præsumtion* of the *Future*, contracted from the *Experience* [of] time *Past*"—but conjectures of the future here are weak as "grounded only upon Experience."[77]

It is unlikely that any one today would think it possible to purge words of their inherent ambiguities or base a political order on such imagined purging. Nor would science bear the certainties it once did for Hobbes. Its claims to superiority over experience would not go unchallenged either, because it is precisely the "experience"—vicarious and direct—of the impact of our science and technology on the biosphere that tells us about their fallibility and reinforces the profound ambivalence that many humans have justly felt about their powers for a very long time. Yet, on the other hand, who would deny that science and technology have been central to human flourishing, both in our numbers and in the quality of the lives that a large number of humans enjoy today? For good or bad—maybe for good *and* bad—we inhabit a world that keeps our brains stimulated far more than was possible at any other period in human history, thanks precisely to our technical inventions. And this same flourishing of the period of the "great acceleration" of human economy and numbers has also created our current sense of a planetary crisis.

Hobbes's discussion leaves us with a question: Can humans learn from the experience of the Anthropocene so that they can renew modern political thought without assuming—contra Tagore and others—that humans are special in the history of complex life or that they are central to the larger scheme of things or that they are even capable of viewing the whole? Can *Homo sapiens* learn to be *Homo prudens*, whatever the political battles that divide us? I should clarify that a project that seeks to understand human loss of reverence for the world they found themselves in is not a project of renouncing the moral courage that humans have always needed in their struggles against various forms of domination and exploitation, including those that have been made possible by modern humans' profligate use of fossil fuels. Indeed, such moral courage can easily coexist with the spirit of reverence whose loss Glikson laments. In many ways, a person like Mohandas Gandhi embodied in himself these two values.[78] Learning to be *Homo prudens* is not to lose moral courage. It is the opposite—to have the moral courage, the courage of a Rachel Carson, for instance—to learn from human experience and to question dominant visions of the human.

The dominant visions of the human cannot any longer be separated from the intensive growth of extractive and consumerist global capitalism that both revealed and exposed humans to the cold indifference of what I have called the planetary. Humans now find themselves in a crisis of temporal management as well, for the planetary calendar (that the IPCC speaks of) and that of the globe (that the UN tries to manage) cannot always be synchronized. I have also suggested that this is in fact a profoundly phenomenological challenge to humans, for we are out-scaled by some of the problems we face. Our embodied selves and our institutions did not evolve to deal with problems that could span geological scales of time. We evolved by taking the work of deep time as given and, over the last couple of hundred years, became "modern" by coming to think of the world around us as something that existed "for us." There were and there still are groups of humans who do not share that assumption, but they lost the battle for the world to the "moderns." The assumption that the world is there mainly to provide a background for the drama of human history to unfold has met with a rude shock (dramatically illustrated by the massive and tragic Australian fires of 2019–2020). We now increasingly know that the planet was not made with humans in view. As we are forced to deal with the planet through dealing with issues like climate change, species extinction, sea level rise, acidification of the seas, extreme weather events, water and food security, and so forth, one hears basically two kinds of calls. They both issue from a recognition of the contemporary human predicament. One is the call to extend human domination of the planet and to make this an "intelligent" planet by ensuring that even where there was once "nature" (by human reckoning) only technology and human justice prevail.[79] This is to continue and intensify the work of the global and to try and fold the planetary back into its reach. The other call—presently utopian but absolutely critical in my judgment and voiced by many cited in this book—is to work toward a planet that no longer belongs to the human-dominant order that European empires, postcolonial and modernizing nationalisms, and capitalist and consumerist globalization created over the last five hundred years, the pace of events accelerating after the 1950s. However humans proceed to deal with this situation—most likely through different mixtures of both options at different levels of organization—they will have to draw their bearings from two connected but different conceptual entities—the globe and the planet—with the former remaining a humanocentric construction and the latter decentering the human in our narratives of the world.

In his thoughtful book, *Learning to Die in the Anthropocene*, Roy Scranton proposed that we prepare ourselves for the death of this consumerist and capitalist civilization.[80] The death of *this* civilization, however, does not mean the death of all possible ideas of civilization. Any just vision of a civilized human future, it seems to me, would have to embrace three principles: (a) all human lives would need to be protected and their flourishing enabled and ensured; (b) biodiversity—which makes for a habitable planet—would have to be protected; and (c) processes of withdrawal from the current human-dominant order of the earth would need to be initiated and advanced. In other words, the humanocentric idea of sustainability will have to speak to the planet-centric idea of habitability. For if my proposition that the intensification of the global has made us encounter the planet is true, then the age of the *purely* global that European empires and capitalism created and that theorists have pondered and historians documented and analyzed since the 1990s is now over. We live on the cusp of the global and the planetary.

* POSTSCRIPT *

The Global Reveals the Planetary

A CONVERSATION WITH BRUNO LATOUR

BRUNO LATOUR: I want to ask you about how to orient ourselves among the planetary conflicts. Actually, there are many different notions of the planet according to you, different ways to feel or become conscious of the planetary dimension of politics. So, I'd like to ask you first about the views associated with the current conservative revolution, and whenever we mention this term a few minutes later Heidegger comes in. How would you define this type of Earth, which could be called the Old Earth or the Reactionary Earth, for then we will be in a better position to locate the others vis-à-vis what you call "the Emergence of the Planetary."

DIPESH CHAKRABARTY: Heuristically, to simplify the story, if I start with the history of labor and capitalism and connect Earth and planet to that, then you will see that in most European languages labor, etymologically, has to do with toil. It's really the physical, wearisome labor of a physical body. It could be a horse. It could be a child. The German word *arbeiten* is etymologically related to the Indo-European word for the (hardworking) "orphan." When Karl Marx is analyzing capitalism, his critical equation is between the machine and man, and even animals. In the first volume of *Capital*, he describes how some of the heavy machines were incorporating movements of a horse's legs and then incorporating movements of human arms and other body parts.[1] That's why Marx says, quoting Goethe, that the machine is robbing the worker of his body. The word *work*, however, is etymologically connected to the word for energy in Greek. This is the seventeenth-century Newtonian definition of "work"—it is energy spent. I think capitalism begins its history with labor as toil. But it discovers, with the development of technology, that it does not need bodily labor to get work done. It can get work done by a waterfall. It can get work done by wind. It can get work done by a machine, by artificial intelligence. The domain of work is what expands under capitalism. In the last few decades with the use of AI, we can see that the future of work that was labor/toil is uncertain. This is giving rise to debates about a universal basic income. What will you do with the people who will not have paid employment? Only some people will have labor that is paid work. As work expands, the reach of capital expands, and our demand on the biosphere expands. This directly affects the biosphere, and it affects deep earth because of resource mobilization.

BL: The extraction of things.

DC: That's right. Extractive capitalism becomes more and more dependent on the biosphere. The more capitalism gets into deep earth through the realm of work, the more it encounters what I call the planet.

BL: Is biosphere a good term here?

DC: The biosphere is part of the critical zone. So, capitalism is making more demands on the critical zone as well as on deep earth.

BL: So it's labor that allows us to discover the planet? Or is it its fabulous extension through capitalism?

DC: It's really the reduction in the importance of labor and the increasing importance of work. This is what makes Marx somewhat obsolete because all of his notions of value, abstract labor, living labor, are based on the presence of human beings, whereas work does not require human presence to the same degree. I can make a mountain do the work for me. This is really the principle of leverage. As capitalism expands it creates a crisis here. The crisis is often what we call in sociology the problem of the future of work. When they say what will happen to the future of work, they mean the future of labor, paid toil. But capitalism expands, creating this crisis, and it actually increases the geomorphological role of humans. That is, the way in which we transform the surface of the planet. The fact that in the Anthropocene they say human beings are the biggest earthmoving agents.

BL: Wait! You're going too fast. We haven't seen the Anthropocene yet; so far, we just have a guy toiling in a field.

DC: Becoming redundant.

BL: There is no "planet" yet.

DC: So, we begin with Heideggerian Earth, and we begin with the eighteenth-century discovery of soil chemistry and the notion of sustainability. This is where one level of capitalism is beginning. What one is introducing is the history of what we call technology.

BL: Production basically.

DC: But, in Carl Schmitt's terms, this is what he calls in English the story of "unencumbered technology" in that book you referred me to.[2]

Amazing, yes? He sees the ship as unencumbered technology because life on a ship depends completely on technology in the ship. This technology, in his terms, in unencumbered because it is not embedded in society as technology on land may be. But technology increasingly gets even more unencumbered, and the more unencumbered technology gets the more you can expand the realm of work.

BL: Until you find that the biosphere has a limit.

DC: Yes, but at the same time, this allows you to come across deep earth, which is part of what I am calling the planet. Work on this level can lead to more earthquakes. Consider this: the use of fossil fuels results in the emission of CO_2. To find fossil fuels, you have to dig deep and need sophisticated technology and machinery to get the work done. The heating of the surface of the planet contributes to geophysical events like earthquakes and tsunamis. There is a book called *Waking the Giant* by the geologist Bill McGuire. Its subtitle is *How a Changing Climate Triggers Earthquakes, Tsunamis, and Volcanoes*.[3]

BL: Is this the movement to the global?

DC: This is where the planetary comes into the global.

BL: Okay. Because you said the global reveals the planetary.

DC: The global discloses the planetary.

BL: Could we get some dates on that?

DC: I would say, the "earth" in Heideggerian terms is older.

BL: In its primordial range.

DC: Yes. The global from the fifteenth century onward. I would say that the planetary begins with nitrogen fixing, the Haber-Bosch process, at the beginning of the twentieth century, because you are actually getting into planetary processes. Of course, the planetary is what the earth comes from. The global comes from a historical process that includes European expansion and the development of a technology that can make the sphere we live on into a globe for us.

BL: Yeah. Without the global we would not have discovered the planetary. And then the planetary, retrospectively, covers billions of years.

DC: Exactly. That's where the historical vista opens up.

BL: So, the planet is antecedent to the global.

DC: Right.

BL: Except it arrives very late, of course.

DC: Yes. We become aware of its presence late.

BL: OK.

DC: What I was going to say here is this. I have two photos—I haven't got them with me, but they are very telling—I juxtaposed two photos of a child in my neighborhood, a four-year-old boy—both walking past an earthmoving machine with no self-awareness and the next minute sitting in a sand pit playing with miniature earthmoving machines moving sand. I use these images to say that the Anthropocene, or humankind's earthmoving agency, has been naturalized to a degree that a little boy is growing up using this toy machinery thinking that this is what humans do. In that sense a planetary role of humans can actually show up in an individual's life, their biography. This child is a colleague's child. One day I walked into their house and saw these toys. I exclaimed to my colleague, "These are Anthropocene toys!" So my colleague, this child's father, became interested in the issue and sent me these photos.

BL: But wait, when this kid is playing it was very positive. Today we might come to a situation when you have the same earthmoving toys but ecologically minded parents might slap kids round the face and say, "Don't play with those horrible things!"

DC: Because we did not call them Anthropocene toys. We called them development toys.

BL: When does the word designate or shift from development to Anthropocene? When does the great acceleration, instead of being positive, become horrifying?

DC: I would say it begins—the awareness (not to be Hegelian about it)—so to talk about beginning of an awareness is not the right term, but surely from Rachel Carson to *The Limits to Growth*, so 1962 to 1972.[4] It's in that decade that a certain shift happens. The shift does not reach Asia. China actually launches the "Four Modernizations" movements in 1978. India liberalizes in 1991 and thinks it's modernizing. The shift is mostly in the West, but I would say that's when there are real doubts. The kind of fight that Rachel Carson had to carry on against the authorities for not being a proper scientist.

BL: So, the planetary emerges as the feeling that there is a clash between the global—basically modernization—and the planet. It's very differentiated in terms of history with every nation being different.

DC: More differentiated than globalization, because the development story runs concurrently with imperialism. The empires say they are developing you. Rostow's stages of growth in the 1950s, specifically 1958. I grew up in India thinking we are organizing development/modernization. To modernize was our ambition. If you go back to the 1950s, the popular defense of technology such as big dams was in terms of providing food to people.

BL: So, the Green Revolution?

DC: Well that's later, 1968 or 1969, but even in the 1950s, dams for instance. The justification of dams was to feed people. From the 1930s population specialists are meeting to ask if there's enough food. Is

Earth's carrying capacity enough? One of the interesting things I see is that the modernization narrative carries with it a certain kind of secular ethic—it's a secular and nonreligious way of caring for the poor. The rhetoric that this will eventually feed people is there both in what right-wing and left-wing people write.

BL: So here there is a beginning of a type of planet at least, it's not yet "the planetary"—we reserve "the planetary" for the Anthropocene?

DC: Yeah.

BL: There is already the feeling of a planet here, but it's a planet as a background and a resource.

DC: Right.

BL: And the question is "how big is the resource?"

DC: And that's how the limit question arises. So, the finitude question.

BL: It's still in development. It's still the global. So, the global has a planet.

DC: The global is shadowed by its own planet.

BL: And then suddenly in the 1970s/1980s in the Western part, the planet is felt both as a resource and that the planetary is somewhat negative or puzzling.

DC: It was revealing itself to James Lovelock and his colleagues in the 1960s, but we were not aware of it.

BL: But the connection with Gaia, according to you, is not as strong as with Earth System Science [ESS].

DC: This is what Gaia mutates into.

BL: And then there is a conflict between the two. But I want to go back to one of the other traits of the different planetary definitions, and that is the question of agency, because, after all, this is from where you started your interest in the climate issue. Clearly the agent [of] history pushing the global was industry.

DC: And for Marx, labor.

BL: And, of course, that's what provided the possibility of socialism. I am sorry—this is a question which you have answered many times: What is the new agent of history now with the planetary? Who is the agent? Is this an agent that is recognizable from the history of socialism? Is it a completely new type of human?

DC: My understanding of the situation is as follows. From the 1970s on something was happening to the figure of the "agent"—every human being as an autonomous agent—which had been so popular in the democratizing, postimperial West of the 1960s. Increasingly, Western complex societies were dealing with the question of agency by compartmentalizing it in different spheres of life. If you were involved in something that could be treated by law, where you would be held responsible or culpable for something, then most law would still use a Lockean notion of personhood. You would be seen as a person with autonomy, you could be held accountable. If you had mental insanity, then you would be forgiven. On the other hand, when doctors discovered that an ulcer was caused by bacteria and not stress, which was late in the 1970s/1980s [in the mid-1980s], a doctor treating you would not treat you as a Lockean person. The doctor would actually say—

BL: You have been infected—

DC: You have a microbiome; there are other living things in you. In a way, this is the world that you [BL] and others laid out, which is a world of connectivity and what you call distributed forms of agency. The world that you and others have talked about, that world was actually increasingly becoming visible, but political institutions, legal institutions were carrying on as before.

BL: With Lockean subjects.

DC: As though that knowledge were not of any political relevance. I think that is the problem we are now facing on a much more increased scale.

BL: So, we no longer have Lockean subjects anywhere?

DC: But we have to pretend that we are still Lockean subjects. That's how we vote. That's how nations talk to each other. In the United Nations every nation has a legal, political personality. When we say big nations are responsible, or when we say capitalism is responsible for greenhouse gas emissions, we speak of capitalism as if it had a moral/legal personality.

BL: So, there is a disconnect between the recognized agents of the past and the new agents of the planetary?

DC: Well, there is a real disconnect—and we don't yet have political bridges to connect the divide.

BL: So, the agent of history is not a human but "earthbounds."

DC: They are earthbound, they are more than human, and political thought has not tried to build any links between humans and these larger complexes of which humans are also a part. We don't yet know how to politically recognize this agent that is more than human and necessarily includes the human.

BL: Would you be comfortable with the adjective *terrestrial*?

DC: Yes.

BL: Not a Lockean subject but terrestrial, an adjective that doesn't specify if they are human or not human.

DC: Exactly. But to view those as subjects.

BL: They are not there.

DC: They are not there, and you have to resignify the word *subject*.

BL: Maybe we can now go through the different things with this argument about agency. The earth—the primordial earth which has obsessed Husserl, Heidegger, and in some way Schmitt and many other Germans—what would have been the human agent there? What would have been the political construction—the agent of history?

DC: Heidegger, if you just remember in some sentences—I forget which essay—where he is comparing, where he develops his distinction between two modes of relating to the earth: demanding from the earth—the German word is *herausfordern*—and leaving something to the mercy of the earth. He says when a peasant sows a seed, he is leaving it to the mercy of the earth, but when I use artificial fertilizers I work the land hard. It is as though I was being demanding of the earth, like holding a gun to somebody's head to rob them. Heidegger clearly favors the position of leaving yourself at the mercy of the earth.

BL: Which is not so very far from the agent who lives in the planetary after many Hegelian twists and turns.

DC: Yes, the agent that discovers that he/she is at the mercy of it anyway.

BL: So that is your point for you when you are accused of having abandoned the revolutionary emancipatory agent of history and when you are talking so much about the new planetary, I mean does it mean that you sort of . . . ?

DC: Here is the problem that I find myself pondering, and I have to think my way through it. When you say, "we have never been modern," you are right in your own terms, in the way you define the constitution of modern. In that sense you are right. But when you finish the book you end with saying you want to keep two things from the constitution of the moderns. One of which is the proliferation of hybrids that humans have produced and that capacity to proliferate hybrids. In other words, even when you are finishing that book—and you are saying we have never been modern in your own terms, which I think is right—you are not saying we should not have MRI machines, should not have chemotherapy, you are not saying that. So that's why I read into that desire to keep that element from the constitution.

BL: But there is an earthiness in your definition of the planetary.

DC: Yes, let me finish this point and then I will come to it. So here is the paradox of human-political thought—and I'll try to say it in a sentence. Political thought from the seventeenth century is founded in the assumption that the role of the state is to provide security of life and property, but in search of making greater numbers of people live safer lives for much longer than previously, we have actually made life more uncertain. The pursuit of that safety has produced now a zone, a very unsafe zone, for human beings. At the same time, the commitment at the level of individual life is built into the goals of all our institutions. If my wife has cancer, I have cancer, we go to the hospital. We are all committed to extending every individual's life, and, as I said, because modern political thought from its very beginning made the individual the focus, the bearer of life and rights and the recipient of welfare, this focus on the individual meant that there was an indifference to the total number of humans. Irrespective of how many humans there are, we will say all of them must have the same rights. So basically there is an indifference to the biosphere built into political thought.

BL: It was the logical way of thinking until the global.

DC: Until the globe ran into the planet.

BL: Because when most of the people were promoting the idea of modernization for everybody, everybody abandons it.

DC: That's the conservative revolution. So, here is the dilemma, then: I think it's very hard now to build a new idea of politics that does not start from the same premise of security of life. Because I think we are all committed to it.

BL: We want to be protected and defended and secured.

DC: Exactly. And even in Heidegger's understanding of what *heimlich*, what "home" is, we'll always find that being at home also has to do with feeling safe. But at the same time the problem is that the present arrangement of things that we thought would make us safe actually makes things unsafe for us. So how do we then bring together political thought and the kind of writings that you and Jane Bennett and others, the so-called new materialists, have done, this is the task that is open. I don't think we know yet — and this is where I find reading your introduction to the catalog interesting because you're also leaving it open. You're saying, I can't define your critical zone for you. It's for you to find out.

BL: You must have an idea what it could mean to be no longer a Lockean human, but an earthbound, or terrestrial human still interested in protection by some sort of entity we might not want to call the state. We don't know yet.

DC: What is the actual shape of the human-to-be, we don't know. And we also have to make an assumption that for every broad agreement that human beings come to, there will be so many interpretations of that agreement that the agreement will fray as soon as we've come to it, which is Schmitt's point about the pluriverse. I cannot produce a description of the futures humans will come to inhabit, but I do feel that the socioeconomic-technological arrangements we currently have cannot go on indefinitely. Late capitalism has become antipolitical because it cannot provide protection for everybody, and therefore there has to be some kind of recognition of the planetary processes of ESS. Now we come back to Gaia, critical zone, and the question of recognizing these variables.

BL: One way of asking the question again is to ask which one of those planetary views carries the possibility of politics. This is, of course, one of the questions you have worked with a lot, which is the global had a powerful way of attributing a political agency and directions which was development—either capitalism or communism—but it was orienting nonetheless. We knew what to do politically and what it meant to be indignant and build political platforms and fight for them. The earth, the primordial earth we talked about in the beginning very quickly turned reactionary because it was not organized as a resistance to capitalism but as a sort of dream of getting away from it. So now suddenly the global reveals the planetary; there are things that are clearly not political. The ESS—which is a planet, it's just one among many planets, and we study them compared to the rest of the cosmos, which means you can't do much politically with it.

DC: But I think the political follows from the planetary because there is already a conservative right-wing construction of planetary politics around.

BL: Yes, of course. Sorry. I forgot that. It's not reactionary in the earth, the primordial earth, sense. It's reactionary in the hypermodernist sense, direction, to go on global all the way to the planet.

DC: In a way, this question that you are asking—what would be the form of politics–it is not a question being asked in a nowhere space, it is actually being asked in a world in which there are powerful people with money and powerful institutions. There is, for instance, the Harvard physicist David Keith, who has written a book in support of geoengineering and who has been given substantial funding from the Gates Foundation for the development of the technology that would enable us to spread aerosol sulfates in the stratosphere. But as a geologist has pointed out, if you spread aerosol sulfates, and you have to keep spreading them for one hundred years to get some breathing space, for those hundred years the sky will be permanently white because of scattered light!

BL: So, in your definition, this is just the global.

DC: This is Gaia extension.

BL: Which swallows the planetary under the same terms. And it doesn't give any political agency except towards people. It doesn't re-politicize the situation very much.

DC: It doesn't. But on the other hand, there has to be a struggle. But what I am saying is that in many ways, because money/capital is already producing a politics of the planet in terms of extending this logic—

BL: Politics in terms of power distribution. But in terms of agent having—to be empowered, to do something of their own existence, no.

DC: No. Not at all. Not at all. But I think the negative part of the current progressive politics is to fight all of this. There are two interesting things in addition—things we didn't talk about. Think of the time when the news of climate change became news for politicians as distinct from scientists, for example, in 1988 when NASA climate scientist James Hansen testified to the US Senate committee on human-driven global warming, and soon after, in the same year, the UN set up the IPCC [Intergovernmental Panel on Climate Change]. Why did they set up IPCC? They set it up following the success of the Montreal Protocol. So, what happened is that we assumed—this is not the political part of the story—we assumed after the Second World War that the United Nations was the proper form—

BL: Right, business as usual.

DC: —for dealing with all global politics, and the United Nations also assumed that the calendar for working on global politics is infinite.

BL: So, when you say global you mean global politics absorbing the planetary as it arrived at the time?

DC: Yeah, but what I am saying is that the assumption of the UN, like for instance with the Israel/Palestine question, is that there is not a finite calendar in politics. But when the climate crisis broke, it was a clash of two kinds of calendars because the scientists were producing a finite calendar, basically saying if you don't do x by this time, then the consequences will be y. Even the 2°C temperature rise figure, as you know, was a politically negotiated figure of a finite calendar. The climate problem was a problem for which there was actually no governance model. The United Nations was not a governance institution meant for dealing with global problems. It gave us a governance model for global problems. The climate problem was the second planetary problem, the first one being the hole in the ozone layer that we dealt with within the parameters of UN processes.

BL: So, for you, when did the planetary emerge even inside the United Nations' format as the impossible to solve question? Is it just because of the inefficiency?

DC: It emerges very gradually, and it shows up very clearly in the

Paris Climate Deal of 2015, where the assumption is that even if everybody met their targets, we would not be able to avoid dangerous climate change unless we produced technology for drawing down greenhouse gases from the air, the technology for which doesn't yet exist. You begin to see that all the nations are trying to act as if they were still on the same global calendar. I can bargain for [some] time here, and some time there. This actually shows that there is a deep crisis of governance. The climate crisis has brought the planet into view, but we don't have a planetary form of governance. Geoengineering and all of these things are taking the place of that politics. On the ground there is already an argument for not mounting global, but local, heterogeneous, multifarious forms of resistance to actions based on the "good Anthropocene" argument.

BL: In the characteristic of your planetary we have to factor in the rhythm of history. So, the primordial earth by definition is always the same, the same village, church bell, there is no history — of the global we know the rhythm: it had to move forward, and we had this expression which sounds strange today, "the acceleration of history" —

DC: And you actually bring in Fernand Braudel here — Braudel is fascinating because he has a complete lack of faith in the individual as the agent of history. In his book *On History*, he actually says that the individual "is all too often a mere abstraction."[5] There are all these big things, but the big things are all very stable. But they are not.

BL: So the acceleration of history is for geology but not for —

DC: Exactly. That is how he seems to have thought. And it did not matter for humans. Recently, I was thinking that you could go and read all the aphorisms of Wittgenstein in *On Certainty* as actually his way to think about what is it that human beings take for granted. One question he asks, if you remember, is this: If you see a building, you ask how old is it, but if you see a mountain, you never ask how old the mountain is. Why?

BL: That's strange.

DC: And that's because the mountain is part of the givenness of the world. But today, reading about the glaciers and climate crisis in South Asia, you realize that the Himalayas are a young mountain range. Australian coal is close to the surface because the mountains are all old and eroded. Our "certainties" are now being shaken up by what I call the percolation of a geological consciousness into our sense of history. That's the tectonic shift happening in history.

BL: But we have been working a lot on that and the question I am trying to ask in addition to it is which one is superseding the others?

DC: The global and the planet?

BL: Well, clearly the global, if we follow your argument about the geoengineering being the global trying to absorb the planetary.

DC: Same for solar energy. Same thing. It's trying to absorb the planetary into the global.

BL: And then there are other versions which I would call more the Gaia line, which is to say we cannot—the planetary has always been there, and it is unique. It has a uniqueness to it, so ESS is the same everywhere.

DC: Gaia is ours. It's unique.

BL: Gaia is singular. We dive into its singularity more and more.

DC: This is very interesting. You are opposing it to astrobiology and the search for exoplanets. The assumption in astrobiology is that we are not singular. That there will be a series one day. And every Gaia will be uniquely bizarre. The planets will be similar, but if there is even life on another planet it's Gaia will be uniquely bizarre.

BL: The argument of Timothy Lenton and Sébastien Dutreuil is that looking for life on other planets is one thing, but looking for Gaia is an absurdity. It is another way to ask the question about the normativity because if we begin to feel that and are interested in the historicity, we learn a new historicity which the global didn't have. This historicity asks to give us also some sort of normativity. It's no longer like what it was when we were supposed to be natural because when we were supposed to be natural . . .

DC: I accept the description, that is why in discussing Gaia and ESS, I actually say ESS is haunted by a poetic intuition, the moment of Gaia. Maybe that is the reason why Lenton and you bring Gaia back in. The point about its singularity is precisely the point about its poetry.

BL: It's not nature.

DC: It's all this poetry. It's not an object of science that repeats itself. In that sense, it is about singularity. The problem is when we as human beings learn about it, and there's no question that there is a sense of the miraculous in knowing about it—I mean that this planet has supported, for one-eighth of its life, this sudden explosion of life-forms, and there is something miraculous about it. I think that's why it goes back to your Facing Gaia lectures. It's very hard to keep that moment of sensing a miracle from some sense of the religious. I make a distinction between wonderment and reverence. I found a geologist who says that humans have lost a sense of reverence for the planet. The Latin root of *reverence* suggests that it means respect with fear, some kind of a feeling that this is much bigger than I am. I have been reading Rudolf Otto on this question.

BL: We are back to a very old idea of nature being terrifying.

DC: When I was growing up in Calcutta—this so-called modern city, the British had built it, but it was already 250 years old when I was born. In my childhood, I was still scared of wild animals in the city—foxes, snakes. As I grew up, the foxes went, the snakes went, the weird frogs—they all went, and today's child is not fearful of wild animals. I opened Adorno and Horkheimer's *Dialectic of Enlightenment*, and the first sentence says a major aim of the Enlightenment was to help humans overcome fear.[6] I became very interested in the question of when did political thought incorporate into itself this overcoming of fear. I was reading Hobbes's *De cive*, which is from 1642, and I realized Hobbes included wild animals in the state of nature. For a historian there is, then, this one task. I agree that modernity has been about overcoming fears of all kinds. We can recall it in a nostalgic way, but as historians we can also write the history of how we came to overcome different kinds of fear because it didn't happen all over the world at the same time. I grew up in a place where fear was very much still a part of my life. Something about that reverence has to be brought back to supplement our very Aristotelian sense of wonderment at the miracle of biodiversity.

Acknowledgments

The debts I ran up in writing this book are too many to count. I mention some and crave the forgiveness of those whose generosity is not formally remembered here. Some are not even nameable! How do I name the gentleman in Düsseldorf who, after hearing me lecture, came up to me in the washroom and recommended, without introducing himself, that I read Eugene Thacker, and left! I did, and I am enormously grateful but don't know the name of my benefactor. Charles Bonner, whom I did not know but who wrote me an old-style letter in 2009 after reading my essay "The Climate of History" introduced me to the work of Reiner Schurmann, whom, much to my shame, I had not read before. There are many other such instances that I can and cannot recall.

Looking back on the matter of being indebted, however, my first debt in writing this book is to some Australian friends—the late Meredith Borthwick, Stephen Henningham, John Hannoush, Roger Stuart, Robin Jeffrey (an honorary Australian once), Katherine Gibson, Sally Hone, Julie Stephens, Marina Bollinger, and Fiona Nicoll—who, during my years of doctoral studies and after, passed on to me their love of their land, its light, the bush, the beach, and the sea that girds their continental nation. Firestorms that destroyed large parts of Canberra and the bush around it in 2003 motivated my interests in the phenomenon of climate change. As I was finishing this book, Australia was burning again. But this time it looked as if the whole continent was on fire. The work on this book began and ended with a sense of bereavement, but not before my engagement with the scientific literature on climate change and Earth System Science had profoundly challenged my humanistic view of humans and their pasts. *The Climate of History in a Planetary Age* is a response to that challenge. But without the love of an unfamiliar land that I inherited as an immigrant, this book would not exist.

The debts I owe to some friends whose work had a direct intellectual influence on this book have to be acknowledged up front. Jan Zalasiewicz, the gifted earth scientist who serves as the chair of the Anthropocene Working Group in London, gave unstintingly of his time and advice whenever I needed them. Bruno Latour, one of the most imaginative and challenging intellectuals of our times, remained interested and involved in the project right from the beginning. François Hartog, the eminent historian and philosopher of historical time, became another treasured friend as the work on this project unfolded. My colleague Bill Brown's deep interest in and explorations of materiality were a source of inspiration and sustenance for this project—perhaps more than he knows. I also recall with pleasure the many conversations and the correspondence I had with Clive Hamilton when he was working on his book *Defiant Earth*. Ewa Domanska remained engaged with the project throughout, giving advice, encouragement, comments, criticisms, and suggestions in equal measure. I completed the first draft of this book, especially of the last chapter, in September 2019, when I was a visitor at the École normale supérieure, Paris, at the invitation of Frédéric Worms. This turned out to be a magical month, not least because of the spontaneous warmth with which Marc Mézard, the director of the École, and Annie Cohen-Solal, his wife and the preeminent biographer of Sartre, took Rochona and me into the loving embrace of their social and intellectual life, and the many stimulating conversations I had in that month with them and with others at the École and in Paris, especially with Frédéric Worms, Pierre Charbonnier, Christophe Bonneuil, Christophe Bouton, Hannes Bajohr (in Berlin), and, of course, with Hartog and Latour. A special word of thanks also to Annebelle Milleville, without whose skills of organization our stay in Paris simply would not have been the transformative experience it turned out to be. I am also deeply grateful to the earth scientist Andrew Glikson of Canberra for sharing with me his thoughts and writings on the Anthropocene.

Nearer home, I have to acknowledge with great pleasure the way gifted and caring colleagues involved in running that wonderful journal of the humanities, *Critical Inquiry*, nurtured and sustained this project from the moment of its very inception. Their comments and criticisms and the stimulation of their own work provided the best nurturing environment I could have asked for. I offer my deepest thanks (in no particular order) to Bill Brown, Tom Mitchell, Laurent Berlant, Hank Scotch, Jay Williams, Françoise Meltzer, Richard Neer, Frances Ferguson, Daniel Morgan, Patrick Jagoda, Haun Saussy, Orit Bashkin, Elizabeth Helsinger, and Heather Keenleyside. Colleagues at the Chicago Center for Contemporary Theory—especially Lisa Wedeen, William Mazzarella,

Bill Sewell, the late Moishe Postone, Jennifer Pitts, Shannon Dawdy, Jo Masco, Kaushik Sunder Rajan, and others—were equally supportive of the project. Two of my busiest colleagues, James Chandler and David Nirenberg, took time out to read and comment on chapter drafts. And there was, besides, my very dear "climate group" at the university—Fredrik Albritton Jonsson, Emily Osborn, Benjamin Morgan, and Julia Adeney Thomas (from the University of Notre Dame)—who regularly organized readings and seminars on the subject, supported by the university's Neubauer Collegium. Fredrik has been my comrade-in-arms in the department of history and beyond. My pleasure in acknowledging what I owe to these individuals is deep, but my debt is deeper. To these friends and to my two deans and the two chairs, my fantastic colleagues and staff of the two departments I belong to—History and South Asian Languages and Civilizations—a big "thank you" for their support and collegiality over the last quarter of a century.

Friends at the Australian National University—Assa Doron, Kirin Narayan, Kenneth George, Meera Ashar, Margaret Jolly, Shameem Black, Chiragh Kasbekar, Will Steffen, Libby Robin, Tom Griffiths, Kuntala Lahiri-Dutt, Fiona Jenkins, and Iain McCalman and Debjani Ganguly, before the latter two left ANU—and at the University of Technology, Sydney—Devleena Ghosh, James Goodman, Tom Morton, Jonathan Marshall, Linda Connor, Stuart Rosenwarne, Ilaria Vanni—provided nothing short of a home away from home for this project. I cannot adequately express my gratitude to them for the many conversations we have had for a couple of decades. I also thank the authorities and the staff of these institutions for their consistent support.

It is also a pleasure to acknowledge some other individuals and institutions that have hosted this project and influenced its development. A fellowship from 2007 to 2008 with Wissenschaftskolleg zu Berlin helped me to find some great conversation partners, especially in Eva Illouz, Michel Chouli, Andrea Büchler, Kristopher König, Axel Meyer, and James Mullet. Starting in 2008, Bernd Scherer has involved me in several events concerning the Anthropocene at the House of World Cultures in Berlin. Parts of this work were presented as the 2013 annual lecture at the Institut für die Wissenschaften vom Menschen, Vienna, at the invitation of the late Krzysztof Michalski and Klaus Nellen. Gary Tomlinson and his colleagues at Yale University invited me to give the Tanner Lectures in Human Values (2014–2015), where Michael Warner, Wai Chee Dimock, Daniel Lord Smail, and Gary himself served as terrific interlocutors. Thomas Lekan of the University of South Carolina invited me in 2015 to be a Provostial Fellow for a week and, along with Robert Emmett of the Rachel Carson Center at Munich, organized a generative workshop on my first essay, "The Climate of History." A visit to the College of the Atlantic in 2016 by invitation from Netta van Vliet and Sarah Hall resulted in the most instructive, engaged, and long-lasting conversation about climate, earth science, and humanities. Ethan Kleinberg invited me to give the 2017 "History and Theory" annual lecture that became the basis for chapter 7. I also include here some material I presented as part of the 2017 Mandel lectures at Brandies University that I gave at the invitation of Ramie Targoff. Elisabeth Décultot invited me to give the inaugural Halle lecture at the University of Halle in 2018, and the conversations I had there with her, Christian Helmreich, Daniel Cyranka, and others remain memorable. I also benefited from the discussions that followed my 2018 annual lecture of the Collegium of Advanced Studies, University of Helsinki. Sections of the last chapter were presented as my 2019 William James lecture at Harvard Divinity School. I hope that Charles Hallisey, Janet Gyatso, and David Lamberth will see how what I learned from them on that visit found its way into this book. An invitation from Hortense Spillers in 2018 to speak at a conference on African and African American studies taught me much on the relationship between race and the anthropocenic world we increasingly inhabit.

I also gained much from the invitations I received to present this work at various other institutions: the Center for Policy Research and the Jawaharlal Nehru University in Delhi; Mahindra Center for the Humanities at Harvard University; the Ludwig-Maximilians-University in Munich; Leiden University; Queen's University, Canada; the University of Toronto; Info-Clio, Switzerland; the University of Wrocław, Poland; Rice University, Houston; Tate Modern, London; Bangladesh History Association; the University of Sydney; Presidency University, Calcutta; the University of Calcutta; Düsseldorf Museum, Germany; the "Anthropocene Campus" run by the Center for the History of Science and Technology at the University of Lisbon, Portugal; Mahatma Gandhi University, Kottayam, India; the humanities program at the University of Pennsylvania; Reed College, Oregon; the University of California at Berkeley and at Los Angeles; Stanford University; the Kolkata Literary Festival, Calcutta; and the organizers of the Samar Sen and Pranabesh Sen Memorial Lectures in Calcutta.

It is also a pleasure to acknowledge some steadfast and old friends who engaged this work at various stages of its development. Homi K. Bhabha, Sanjay Seth, Rajyasree Pandey, Saurabh Dube, Ajay Skaria, Sheldon Pollock, Arjun Appadurai, Amitav Ghosh, Partha Chatterjee, Gautam Bhadra, and Shahid Amin have remained sparring partners, reading, debating, and discussing with me many aspects of my various projects. Sabyasachi Bhattacharya and Soumya Chakravarti, two physicists whose friendship feels ancient and strong like an oak tree, have acted as my bridge to the physical sciences. Many of my former and current students have warmly supported this work through their conversations in and outside the classroom. I need to make special mention, however, of Nazmul Sultan, who led me to some excellent literature on Arendt and with whom I had many discussions on the contemporary state of theory in the humanities and the social sciences. I am grateful for the encouragement and comments I received from various other colleagues, among whom the following come to mind immediately: Navroz Dubash, Liz Chatterjee, Emma Rothschild, Joyce Chaplin, Stephen Greenblatt, Sheila Jasanoff, Sverker Sörlin, Neil Brenner, Stephen Muecke, Norman Wirzba, Awadhendra Saran, Sebastian Conrad, Andreas Eckert, Uma Das Gupta, Henning Trüper, Kunal and Shubhra Chakrabarti, Daniel and Ellen Eisenberg, Prathama Banerjee, Dominic Boyer, Cymene Howe, Timothy Morton, Eric Santner, Robert Pippin, Raghuram Rajan, Dilip Gaonkar, Chandi Prasad Nanda, Judit Carrera, Peter Wagner, Thomas Blom Hansen, Aniket De, Barnhard Malkmus, Raghabendra Chattopadhyay, Bo Stråth, Rita Brara, Sunil Amrith, Eva Horn, Helge Jordheim, Franz Mauelshagen, Shruti Kapila, Faisal Devji, Miranda Johnson, Arvind Elangovan, Arnab Dey, Dwaipayan Sen, Minal Pathak, and Sumit and Tanika Sarkar. It was a delight to meet Samuel Garrett Zeitlin, who turned out to be a superb guide to reading Schmitt. Ian Baucom has always taken a generous and deep interest in this project. I am sorry that both Ian's new book, *History 4° Celsius: Search for a Method in the Age of the Anthropocene* (Duke University Press, 2020), offering a powerful critique of my work, and Achille Mbembe's recent writings on planetarity came too late into my hands for me to make use of. I also recall with gratitude and sadness three friends who were interested in this book but did not live to see it: Christopher Bayly, Don Willard, and Raghab Bandyopadhyay. Gerard Siarny, a friend for a long time and my research assistant for the last several years, has always helped with my research, prose, presentation, and argument. Many, many thanks to him.

Alan Thomas, my editor and friend, has been the model of patience and understanding, allowing this project time to take its own shape. At the same time, his sage editorial advice has saved me from many errors. I am deeply grateful to him and to his colleagues at the University of Chicago Press, especially Randolph Petilos, for their interest in this work and the support I have received from them. I am also grateful to the press's two anonymous readers who read a draft manuscript and had useful comments to offer.

And now to turn to some words of thanks that are deeply personal: to Rochona, who bravely fought off cancer from 2018 to 2019 and helped me manage my work and life; to Kaveri, who had her own cancer to battle, and Arko for their interest in this book and in fighting climate change; to Roopa Majumdar, Boria Majumdar, Sharmistha Gooptu-Majumdar, Aisha Gooptu-Majumdar, and Shyamapada Ray, whose presence livens up my life when I am in my beloved and impossible city of Calcutta; to Partha Sen Gupta, Utpalendu Gupta, Ashoka Chatterjee, Subir Chakrabarti, Neptune Srimal, Aditi Mody, Payal Chawla, Debiprakash and Tandra Basu, Rita Chattopadhyay, Raja and Chaitali Dasgupta, Anil Acharya, Semanti and Tridibesh Ghosh, Durba Bandyopadhyay, Shahduzzaman, Chinmoy Guha, Sanjib Mukhopadhyay, Sabbir Azam, and Ahmed Kamal for their encouragement and friendship; to Saptarshi Ghatak who kept me musical company from faraway Calcutta while I recovered from a nasty road accident in 2019 during the months of November and December; to Barbara Willard, Lisa Wedeen and Don Reneau, Neeraj and Meenakshi Jolly, Mohan and Lalitha Gundeti, Bill Brown and Diana Young, James and Elizabeth Chandler, Muzaffar and Rizwana Alam, and our many doctoral students for their presence in our everyday lives; to Khurshid and Peter Roeper, Siddhartha and Chandana De, and Gautam and Ruplekha Biswas who constitute my larger Bengali family in Canberra.

My final thanks are to those whose writings taught me to appreciate the rich and precarious nature of life. With them, and with those mentioned or inadvertently forgotten here, I am glad to have shared this earth, my worlds, a globe, and a planet.

* * *

I am grateful to several editors and publishers for permission to incorporate into this book ideas I first tried out in articles published in their journals and publications:

- A section of the introduction incorporates material published in "Museums between Globalization and the Anthropocene," *Museum International* 71 (2019): 12–19.
- Chapter 1 is an updated and revised version of "The Climate of History: Four Theses," *Critical Inquiry* 35, no. 2 (Winter 2009): 197–222.
- Chapter 2 is a revised version of "Climate and Capital: On Conjoined Histories," *Critical Inquiry* 41, no. 1 (Autumn 2014): 1–23.
- Chapter 3 was originally published as "The Planet: An Emergent Humanist Category?," *Critical Inquiry* 46, no. 1 (Autumn 2019): 1–31.
- Chapter 4 draws upon my essay "Planetary Crisis and the Difficulty

of Being Modern," *Millennium: Journal of International Studies* 46, no. 3 (2018): 1–24.
- Chapter 5 was originally published as "The Dalit Body: A Reading for the Anthropocene," in *The Empire of Disgust: Stigma and the Law*, ed. Martha Nussbaum and Zoya Hasan (Delhi: Oxford University Press, 2018): 1–20.
- The first version of chapter 6 was published as "Humanities in a Warming World: The Crisis of an Enduring Kantian Fable," *New Literary History* 47, no. 2/3 (2016): 377–97.
- Chapter 7 was originally published as "Anthropocene Time," *History and Theory* 57, no. 1 (March 2018): 5–32.
- Some sections of chapter 8 draw from my "The Human Condition in the Anthropocene," in *The Tanner Lectures on Human Values*, vol. 35, ed. Mark Matheson (Salt Lake City: University of Utah Press, 2016), 137–88, and "The Planet: An Emergent Matter of Spiritual Concern?," *Harvard Divinity Bulletin* 47, no. 3/4 (Winter 2019): 28–38.
- The postscript is reprinted from Bruno Latour and Peter Weibel, eds. *Critical Zones: The Science and Politics of Landing on Earth* (Cambridge, MA: MIT Press, 2020).

Chicago
May 26, 2020

Notes

INTRODUCTION

1. Steven Nadler, *Spinoza: A Life* (Cambridge: Cambridge University Press, 2001), 220.

2. See my "An Era of Pandemics? What Is Global and What Is Planetary about COVID-19," *In the Moment* (blog), *Critical Inquiry*, October 16, 2020, https://critinq.wordpress.com/2020/10/16/an-era-of-pandemics-what-is-global-and-what-is-planetary-about-covid-19/.

3. Ken Ruthven, ed., *Beyond the Disciplines: The New Humanities* (Canberra: The Australian Academy of the Humanities, 1992).

4. For two recent stimulating discussions on this theme, see Sebastian Conrad, *What Is Global History?* (Princeton, NJ: Princeton University Press, 2016), and Sumathi Ramaswamy, *Terrestrial Lessons: The Conquest of the World as Globe* (Chicago: University of Chicago Press, 2017).

5. Jan Zalasiewicz et al., "A General Introduction to the Anthropocene," in *The Anthropocene as a Geological Time Unit: A Guide to the Scientific Evidence and Current Debate*, ed. Jan Zalasiewicz, Colin Neil Waters, Mark Williams, and Colin Summerhayes (Cambridge: Cambridge University Press, 2019), 2–11. Four other very important introductions to the problem of the Anthropocene, discussed from the point of view of the humanities and the social sciences, are Simon Lewis and Mark Maslin, *The Human Planet: How We Created the Anthropocene* (London: Penguin Random House, 2018); Jeremy Davies, *The Birth of the Anthropocene* (Berkeley: University of California Press, 2018); Eva Horn and Hannes Bergthaller, *The Anthropocene: Key Issues for the Humanities* (London: Routledge, 2020); and Carolyn Merchant, *The Anthropocene and the Humanities: From Climate Change to a New Age of Sustainability* (New Haven, CT: Yale University Press, 2020).

6. Some of these debates are recounted in my "The Human Significance of the Anthropocene," in *Modernity Reset!*, ed. Bruno Latour (Cambridge, MA: MIT Press, 2016).

7. Zalasiewicz et al., *The Anthropocene*, 31–40, for arguments regarding the utility of the formalization of the term.

8. Cited in Andrew S. Goudie and Heather A. Viles, *Geomorphology in the Anthropocene* (Cambridge: Cambridge University Press, 2016), 28. See also the larger discussion

in J. R. McNeill and Peter Engelke, *The Great Acceleration: An Environmental History of the Anthropocene since 1945* (Cambridge, MA: Harvard University Press, 2014).

9. Peter Haff, "Technology as a Geological Phenomenon: Implications for Human Well-Being," in *A Stratigraphical Basis for the Anthropocene*, ed. C. N. Waters et al. (London: Geological Society, Special Publications, 2014), 301–2.

10. Haff, 302.

11. For a critique of Haff's concept of the technosphere, see Jonathan F. Donges et al., "The Technosphere in Earth System Analysis: A Coevolutionary Perspective," *Anthropocene Review* 4, no. 1 (2017): 23–33.

12. Carl Schmitt, *Dialogues on Power and Space*, ed. Andreas Kalyvas and Frederico Finchelstein, trans. and with an introduction by Samuel Garrett Zeitlin (Cambridge: Polity, 2015; first published in German, 1958), 72, 73–74.

13. Peter Haff, "The Technosphere and Its Relation to the Anthropocene," in Zalasiewicz et al., *Anthropocene as a Geological Time Unit*, 143.

14. Jan Zalasiewicz et al., "Scale and Diversity of the Physical Technosphere: A Geological Perspective," *Anthropocene Review* 4, no. 1 (2017): 16.

15. University of Leicester, "Earth's 'Technosphere' Now Weighs 30 Trillion Tons, Research Finds," Phys.org, November 30, 2016, https://phys.org/news/2016-11-earth-technosphere-trillion-tons.html. The basis for the calculations are presented in Zalasiewicz et al., "Scale and Diversity of the Physical Technosphere," 9–22.

16. Zalasiewicz et al., *Anthropocene as a Geological Time Unit*, 105.

17. Goudie and Viles, *Geomorphology in the Anthropocene*, 33.

18. Zalasiewicz et al., *Anthropocene as a Geological Time Unit*, 71.

19. I am in fundamental agreement with Jeremy Davies's point that the humanities' uptake of the discussion on climate change and the Anthropocene involves the question of what to do—in writing human history or politics—with deep time. Davis, *Birth of the Anthropocene*.

20. Naomi Oreskes, "Scaling Up Our Vision," *Isis* 105 (2014): 388. On the question of extinctions and why they pose a problem for human existence, see the discussion in Peter F. Sale, *Our Dying Planet: An Ecologist's View of the Crisis We Face* (Berkeley: University of California Press, 2011), 102, 148–49, 203–21, 233. See also Elizabeth Kolbert, *The Sixth Extinction: An Unnatural History* (New York: Henry Holt, 2014).

21. Frédéric Worms, *Pour un humanism vital: Lettres sur la vie, la mort, le moment present* (Paris: Odile Jacob, 2019).

22. Here I register—with respect and admiration—a small conceptual disagreement with some of the propositions Daniel Lord Smail has put forward in his thought-provoking book *On Deep History and the Brain* (Berkeley: University of California Press, 2008). The book opens with the statement "If humanity is the proper subject of history, as Linnaeus might well have counseled, then it stands to reason that the Paleolithic era, that long stretch of the Stone Age before the turn to agriculture, is part of our history" (2). I agree, but then Smail goes on to say, with regard to the genes ("of considerable antiquity") that are "responsible for building the autonomic nervous system," that "this history is also world history since the equipment is shared by all humans though it is built, manipulated, and tweaked in different ways by different cultures" (201). True, but the physical feature of the autonomic nervous sys-

tem is something humans share with many other animals, so this could not quite be a world history of humans alone. We should perhaps move toward writing these histories shared between different species, but that is a separate discussion. However, philosopher Catherine Malabou's speculations based on the history of the human brain and its plasticity are highly relevant and, if borne out by future developments, may indeed challenge some of my claims. Catherine Malabou, "The Brain of History, or The Mentality of the Anthropocene," *South Atlantic Quarterly* 116, no. 1 (January 2017): 39–53. See also her *What Should We Do with Our Brain?*, trans. Sebastian Rand (New York: Fordham University Press, 2008).

23. For generative thoughts on how the Anthropocene impinges on our usual understanding of politics and political thought, see Duncan Kelly, *Politics and the Anthropocene* (Cambridge: Polity, 2019).

24. See Patchen Markell's excellent essay "Arendt's Work: On the Architecture of *The Human Condition*," *College Literature* 38, no. 1 (Winter 2011): 36, 37n3, where he comments on the "interdependence" of work and action. Note also the following comment of Vatter's: "Action and natality . . . stand in what may be called a 'mimetic' relation with respect to each: action can only be the intensification of natality, and never its limitation, control or domination." Miguel Vatter, "Natality and Biopolitics in Hannah Arendt," *Revista de ciencia política* 26, no. 2 (2006): 155.

25. Hannah Arendt, *The Human Condition*, 2nd ed., with an introduction by Margaret Canovan (1958; Chicago: University of Chicago Press, 1998), 7, 175 (hereafter HC).

26. Arendt, HC, 8–9. Arendt made the same point in *The Life of the Mind*: "had Kant known of Augustine's philosophy of natality he might have agreed that the freedom of a *relatively* absolute spontaneity is no more embarrassing to human reason that the fact that men are *born*—newcomers again and again *in a world that preceded them in time*." Hannah Arendt, *The Life of the Mind*, 2 vols. (1971; New York: Harcourt, 1978), 2:110, emphasis added. Vatter, "Natality," 137–59, offers a penetrating discussion of the genesis of Arendt's category, "natality," and of its treatment in the literature on Arendt. Arendt emphasized natality presumably to distinguish her thought from the Heideggerian tradition of thinking from the horizon of mortality and finitude. For an extended discussion on this point of whether natality opposes Heidegger's emphasis on finitude and mortality or is derived from Heidegger's treatment of birth, see Vatter, 138–39. Vatter emphasizes the independent derivation of this concept and has some very interesting observations to offer on connections between Walter Benjamin and Arendt on the question of natality. Dana Villa's *Arendt and Heidegger: The Fate of the Political* (Princeton, NJ: Princeton University Press, 1995) contains a helpful discussion of Arendt's appropriation of Heidegger for the purposes of her own philosophy.

27. Arendt, HC, 9.

28. Arendt, *Life of the Mind*, 2:109. See also Hannah Arendt, "Lying in Politics," in her *Crisis of the Republic* (1969; New York: Harcourt, Brace, 1972), 5: "A characteristic of human action is that it always begins something new, and this does not mean that it is ever permitted to start *ab ovo*, to create *ex nihilo*."

29. Hannah Arendt, *The Promise of Politics* (New York: Schocken Books, 2005), 95, cited in Vatter, "Natality," 142.

30. See, for example, Sheldon S. Olin, "Hannah Arendt: Democracy and the Politi-

cal," *Salmagundi*, no. 60 (Spring/Summer 1983): 3–19; Hannah Fenichel Pitkin, "Justice: On Relating Private and Public," *Political Theory* 9, no. 3 (August 1981): 327–52; Keith Breen, "Violence and Power: A Critique of Hannah Arendt on 'the Political,'" *Philosophy and Social Criticism* 33, no. 3 (2007): 343–72; Jacques Rancière, "Ten Theses on Politics," *Theory and Event* 5, no. 3 (2001).

31. See in particular Markell, "Arendt's Work." See also Steven Klein, "'Fit to Enter the World': Hannah Arendt on Politics, Economics, and the Welfare State," *American Political Science Review* 108, no. 4 (November 2014): 856–69.

32. The trouble that commentators have had with these distinctions is summarized in Markell, "Arendt's Work," 15–44. See in particular the discussion on 15–17. Markell also draws some interesting lines of connection between Arendt's 1951 interpretation of totalitarianism as a system that reduced juridical persons to "merely natural beings" and her later development of the distinction between "labor" and "work" (see p. 19).

33. "Work . . . breaks out of the cyclical time of labor: it is the activity through which human beings fabricate a world of *durable* objects." Markell, "Arendt's Work," 22, emphasis added. See also his comments on pp. 27 and 32. Durability, it seems to me, has to extend beyond the present and across generations for "work" to retain the character that Arendt assigned to it.

34. Arendt, *HC*, 137. On the relatedness of work and action, see Markell, "Arendt's Work," 20. Arendt's idea of natality does not have to be read as an argument in favor of birth-related identity. See Vatter, "Natality," 151, 152: "natality is essentially antecedent with respect to the common world." A somewhat similar interpretation of natality in Arendt's writings is available in Peg Birmingham's book *Hannah Arendt and Human Rights: The Predicament of Common Responsibility* (Bloomington: Indiana University Press, 2006), 76: "She [Arendt] is continually preoccupied with the double miracle of the event of natality, both the miracle of the given and the miracle of beginning."

35. Arendt, *HC*, 173.

36. Arendt's discussion (*HC*, 167–74) moves toward the consideration of the relationship between work and art. Markell, "Arendt's Work," provides an arresting commentary on the complications of this move.

37. A powerful if pessimistic treatment of the problem of intergenerational ethics and responsibility in the context of climate change is Stephen Gardiner's *A Perfect Moral Storm: The Ethical Tragedy of Climate Change* (New York: Oxford University Press, 2011).

38. Indian newspaper reports say that 31 out of 128 hills in the Aravali range in the state of Rajasthan have "vanished in the last 50 years due to massive illegal mining." Amit Anand Choudhary, "A Fourth of Aravali Hills in Rajasthan Gone Forever," *Times of India*, October 23, 2018, https://timesofindia.indiatimes.com/india/31-hills-in-aravali-region-in-rajasthan-disappeared-sc-directs-state-to-stop-illegal-mining-in-48-hours/articleshow/66336416.cms.

39. For the debate on geoengineering, see Clive Hamilton, *Earthmasters: The Dawn of the Age of Climate Engineering* (New Haven, CT: Yale University Press, 2013); David Keith, *A Case for Climate Engineering* (Cambridge, MA: MIT Press, 2013); and Holly Jean Buck, *After Geoengineering: Climate Tragedy, Repair, and Restoration* (London: Verso, 2019). For warnings regarding a possible rebarbarization of the world, see Bruno Latour,

Down to Earth: Politics in the New Climatic Regime, trans. Catherine Porter (Cambridge: Polity, 2018; first published in French, 2017).

40. Kelly, *Politics and the Anthropocene*; Geoff Mann and Joel Wainwright, *Climate Leviathan: A Political Theory of Our Planetary Future* (London: Verso, 2018). The problem of temporality described here also gives rise to the idea of "end-time" that has received a very thoughtful treatment in François Hartog, *Chronos: L'Occident aux prises avec le Temps*, Collection Bibliothèque des Histoires (Paris: Gallimard, 2020).

41. Carl Schmitt, *The Concept of the Political*, trans. and with an introduction by George Schwab (Chicago: University of Chicago Press, 1996), 44–45, 53. In this sense, I remain skeptical of the practicalities of propositions that see the solution to capital's insatiable drive for accumulation in some single rational principle that all humans will accept, voluntarily or involuntarily. See John Bellamy Foster, Brett Clark, and Richard York, *The Ecological Rift: Capitalism's War on the Earth* (New York: Monthly Review Press, 2010), 417, 436. Why such a state of affairs, even if achieved through some revolution, could not be gamed by some humans in parochial interest has never been clear to me. Humans, it seems to me, will have to work through the reality of their conflicting interests, desires, power, and imagination.

42. See my *Provincializing Europe: Postcolonial Thought and Historical Difference* (2000; Princeton, NJ: Princeton University Press, 2008), introduction.

43. Zalasiewicz et al., *Anthropocene as Geological Time Unit*, 11.

44. Bernard Williams, *Truth and Truthfulness* (Princeton, NJ: Princeton University Press, 2002), 3. The immediate target of Williams's statement is Bruno Latour. I think Williams is unfair to Latour. But he does have a point, unfortunately. I have come across critical writings in science studies that appear to be laboring under the assumption that Williams criticizes.

45. After all, if science is nothing but politics (however understood), then what is the cognitive status of this statement itself? Is it more certain than science or politics?

46. Anthropologists have often documented the practical-political need that exists in many parts of the world to translate the abstract sciences of climate change and the Anthropocene into locally legible terms and concerns. See, for instance, the discussions in Sara de Wit, "To See or Not to See: On the 'Absence' of Climate Change (Discourse) in Maasailand, Northern Tanzania," and Vimbai Kwashirai, "Perspectives on Climate Change in Makonde District, Zimbabwe since 2000," in *Environmental Change and African Societies*, ed. Ingo Haltermann and Julia Tischler (Leiden: Brill, 2019), 23–47, 48–70.

47. See my essay "Postcolonial Studies and the Challenge of Climate Change" in my book *The Crises of Civilization: Exploring Global and Planetary Histories* (New Delhi: Oxford University Press, 2018), 223–43.

48. See C. B. Macpherson, *The Theory of Possessive Individualism: Hobbes to Locke* (Oxford: Clarendon, 1962).

49. The archaeologist Kathleen D. Morrison states that the "codification of several elite cuisines based on irrigated produce, especially rice" can be documented from "the first millennium C.E. in South India." See her "The Human Face of the Land: Why the Past Matters for India's Environmental Future," NMML Occasional Paper, History and Society, New Series, no. 27 (New Delhi: Nehru Memorial Museum and Library, 2013), 16.

50. "Whose Anthropocene? Revisiting Dipesh Chakrabarty's 'Four Theses,'" ed.

Robert Emmett and Thomas Lekan, special issue, *Transformations in Environment and Society*, no.2 (2016), and my essay "The Politics of Climate Is More Than the Politics of Capitalism," *Theory, Culture, Society* 34, no. 2/3 (2017): 1–13. Relevant feminist criticisms of my position in the original version of the "The Climate of History" essay may be found in Richard Grusin, ed., *Anthropocene Feminism* (Minneapolis: University of Minnesota Press, 2017) — see in particular Grusin's introduction and the essays by Claire Colebrook and Stacy Alaimo. See also Tom Cohen and Claire Colebrook, "Vortices: On 'Critical Climate Change' as a Project," *South Atlantic Quarterly* 116, no. 1 (2017): 129–43. For a recent critique of my position, see Dan Boscov-Ellen, "Whose Universal? Dipesh Chakrabarty and the Anthropocene," *Capitalism, Nature, Socialism* 31, no. 1 (2020): 70–83, published online August 23, 2018, https://www.tandfonline.com/doi/abs/10.1080/10455752.2018.1514060?journalCode=rcns20. The most sustained, generous, and yet resolute critique of my work on climate change so far is to be found in Ian Baucom's *History 4° Celsius: Search for a Method in the Age of the Anthropocene* (Durham, NC: Duke University Press, 2020).

51. "In his . . . 'Climate of History' article of 2009," writes Wood, "Dipesh Chakrabarty argued that the geological scale of climate change alters the baseline terms for critique. . . . Theorizing that climate change demands a species-level category of human agency and a 'deep time' elucidation of world history, Chakrabarty insisted, moving forward, on both fine-grained historical differentiation *and* a planetary, deep-time consciousness, but his critics . . . have caricatured his argument as binary, involving a zero-sum choice between 'critique' and 'science.'" Gillen D'Arcy Wood, "Climate Delusion: Hurricane Sandy, Sea Level Rise, and 1840s Catastrophism," *Humanities* 8, no. 131 (2019): 3–4.

52. https://www.sealevel.info/1988_Hansen_Senate_Testimony.html.

53. Edward Said, *Orientalism* (1978; New York: Vintage, 1994).

54. For an older description of what the humanities were understood to be in the American academic scene, see Irving Babbit, *Literature and the American College: In Defense of the Humanities* (Boston: Houghton, Mifflin, 1908). See also my essay "An Anti-Colonial History of Postcolonial Thought: A Tribute to Greg Dening" in my *Crises of Civilization*, 28–53, and some of the other essays included in that book for a sense of the transformation that the humanities underwent in the second half of the twentieth century.

55. James C. Scott, *Weapons of the Weak: Everyday Forms of Peasant Resistance* (New Haven, CT: Yale University Press, 1985).

56. The first volume of Ranajit Guha, ed., *Subaltern Studies: Writings on Indian History and Society* (New Delhi: Oxford University Press), was published in 1982.

57. The history and ramifications of that essay are recounted in Rosalind C. Morris, ed., *Reflections on the History of an Idea: Can the Subaltern Speak?* (New York: Columbia University Press, 2010).

58. Alan Reed, ed., *The Fact of Blackness: Frantz Fanon and Visual Representation* (London: Institute of Contemporary Arts, and Seattle: Bay Press, 1996).

59. Homi K. Bhabha, *The Location of Culture* (London: Routledge, 1994).

60. Donella Meadows et al., *The Limits to Growth* (New York: Universe Books, 1972).

61. The classic text is Anil Agarwal and Sunita Narain, *Global Warming in an Unequal World: A Case of Environmental Colonialism* (New Delhi: Centre for Science and Environ-

ment, 1991). Rob Nixon's *Slow Violence and the Environmentalism of the Poor* (Cambridge, MA: Harvard University Press, 2011) is justly celebrated for being a text of postcolonial humanities that brings together concerns about both human inequalities and environmental degradation caused by the operations of capitalist enterprises.

62. Sverre Raffnsøe, *Philosophy of the Anthropocene* (Houndmills: Palgrave Macmillan, 2016), 53n1.

63. *Kant's Introduction to Logic and his The Mistaken Subtilty of the Four Figures*, trans. Thomas Kingsmill Abbott "With a Few Notes by Coleridge" (London: Longmans, Green, 1885), 1. These questions stayed with Kant for a long time. See his *Critique of Pure Reason*, trans. Norman Kemp Smith (1929; London: Macmillan, 1980; first published in German, 1781), 635–36 (A805, B833; A806, B834).

CHAPTER ONE

1. Alan Weisman, *The World without Us* (New York: Thomas Dunne, 2007), 3–5.

2. See C. A. Bayly, *The Birth of the Modern World, 1780–1914: Global Connections and Comparisons* (Malden, MA: Blackwell, 2004).

3. The prehistory of the science of global warming going back to nineteenth-century European scientists like Joseph Fourier, Louis Agassiz, and Arrhenius is recounted in many popular publications. See, for example, the book by Bert Bolin, the chairman of the UN's Intergovernmental Panel on Climate Change (1988–1997), *A History of the Science and Politics of Climate Change: The Role of the Intergovernmental Panel on Climate Change* (Cambridge: Cambridge University Press, 2007), pt. 1. See also David Archer and Raymond Pierrehumbert, eds., *The Warming Papers: The Scientific Foundation for the Climate Change Forecast* (Oxford: Wiley-Blackwell, 2011).

4. Quoted in Mark Bowen, *Censoring Science: Inside the Political Attack on Dr. James Hansen and the Truth of Global Warming* (New York: Dutton, 2008), 1.

5. Quoted in Bowen, 228. See also "Too Hot to Handle: Recent Efforts to Censor Jim Hansen," *Boston Globe*, February 5, 2006, E1.

6. See, for example, Walter K. Dodds, *Humanity's Footprint: Momentum, Impact, and Our Global Environment* (New York: Columbia University Press, 2008), 11–62.

7. This statement, when first published, caused much understandable ire among Marxist friends but really was the beginning of my journey toward developing a distinction between the globe and the planet that is so central to the overall argument of this book.

8. Giovanni Arrighi, *The Long Twentieth Century: Money, Power, and the Origins of Our Times* (1994; London: Verso, 2006), 356; *Adam Smith in Beijing: Lineages of the Twenty-First Century* (London: Verso, 2007), 387–89.

9. An indication of the growing popularity of the topic is the number of books published recently with the aim of educating the general reading public about the nature of the crisis. Here are a few titles that originally informed this chapter: Mark Maslin, *Global Warming: A Very Short Introduction* (Oxford: Oxford University Press, 2004); Tim Flannery, *The Weather Makers: The History and Future Impact of Climate Change* (Melbourne: Melbourne Text, 2005); David Archer, *Global Warming: Understanding the Forecast* (Malden, MA: Blackwell, 2007); Kelly Knauer, ed., *Global Warming* (New York: Time, 2007); Mark Lynas, *Six Degrees: Our Future on a Hotter Planet* (Washington, DC: National Geographic, 2008); William H. Calvin, *Global Fever: How to Treat Climate*

Change (Chicago: University of Chicago Press, 2008); James Hansen, "Climate Catastrophe," *New Scientist*, July 28–August 3, 2007, 30–34; James Hansen et al., "Dangerous Human-Made Interference with Climate: A GISS ModelE Study," *Atmospheric Chemistry and Physics* 7, no. 9 (2007): 2287–312; and Hansen et al., "Climate Change and Trace Gases," *Philosophical Transactions of the Royal Society*, July 15, 2007, 1925–54. See also Nicholas Stern, *The Economics of Climate Change: The "Stern Review"* (Cambridge: Cambridge University Press, 2007). References to more recent scientific and other kinds of literature will be found distributed throughout this book.

10. Naomi Oreskes, "The Scientific Consensus on Climate Change: How Do We Know We're Not Wrong?," in *Climate Change: What It Means for Us, Our Children, and Our Grandchildren*, ed. Joseph F. C. Dimento and Pamela Doughman (Cambridge, MA: MIT Press, 2007), 73, 74.

11. The proposition that the current episode of planetary warming is anthropogenic in origin has only gained more ground since I wrote this paragraph. The science now is seldom questioned the way it used to be even six or seven years ago. Significantly, I should add, most if not all of the scientists who have so far taken it upon themselves to explain to the general reader the problem of planetary climate change are from the so-called West. The IPCC, for sure, has scientists participating from all over the world, but most of these scientists make specialist contributions. The scientists I depend on here are all from what we generally regard as the West. This may very well say something about the historical capacity of the once-imperial West to produce and speak on behalf of "universals." It also shows the uneven nature of the public sphere for discussions of global warming. I do not investigate this problem here, but there is no denying that this book comes out of the public sphere around the question of global warming that the Western academy has created.

12. While my point about Vico stands, there was clearly another intellectual trend, prevalent well into the eighteenth century: the Biblical histories of the world that thought of natural and human history as part of the same creation story. The real separation of the two happened in the nineteenth century with the rise of the modern social sciences. See Jean-Baptiste Fressoz and Fabien Locher, "Modernity's Frail Climate: A Climate History of Environmental Reflexivity," *Critical Inquiry* 38, no. 3 (2012): 579–98.

13. A long history of this distinction is traced in Paolo Rossi, *The Dark Abyss of Time: The History of the Earth and the History of Nations from Hooke to Vico*, trans. Lydia G. Cochrane (1979; Chicago: University of Chicago Press, 1984).

14. Benedetto Croce, *The Philosophy of Giambattista Vico*, trans. R. G. Collingwood (1913; New Brunswick, NJ: Transaction, 2002), 5. Carlo Ginzburg has alerted me to problems with Collingwood's translation.

15. See the discussion in Perez Zagorin, "Vico's Theory of Knowledge: A Critique," *Philosophical Quarterly* 34 (January 1984): 15–30.

16. Karl Marx, "The Eighteenth Brumaire of Louis Bonaparte," in Marx and Frederick Engels, *Selected Works* (Moscow: Progress, 1969), 1:398. See V. Gordon Childe, *Man Makes Himself* (London: Watts, 1941). Indeed, Althusser's revolt in the 1960s against humanism in Marx was in part a jihad against the remnants of Vico in the savant's texts; Étienne Balibar, personal communication with the author, December 1,

2007. I am grateful to Étienne Balibar and the late Ian Bedford for drawing my attention to complexities in Marx's connections to Vico.

17. David Roberts describes Collingwood as "the lonely Oxford historicist . . . , in important respects a follower of Croce's." David D. Roberts, *Benedetto Croce and the Uses of Historicism* (Berkeley: University of California Press, 1987), 325.

18. On Croce's misreading of Vico, see the discussion in general in Cecilia Miller, *Giambattista Vico: Imagination and Historical Knowledge* (Basingstoke: Macmillan, 1993), and James C. Morrison, "Vico's Principle of Verum is Factum and the Problem of Historicism," *Journal of the History of Ideas* 39 (October–December 1978): 579–95.

19. R. G. Collingwood, *The Idea of History* (1946; London: Oxford University Press, 1976), 212–14, 216.

20. Collingwood, 193.

21. Roberts, *Benedetto Croce and the Uses of Historicism*, 59, 60, 62.

22. Joseph Stalin, *Dialectical and Historical Materialism* (1938), https://www.marxists.org/reference/archive/stalin/works/1938/09.htm.

23. Fernand Braudel, "Preface to the First Edition," in *The Mediterranean and the Mediterranean World in the Age of Philip II*, trans. Siân Reynolds (1949; London: Collins, 1972), 1:20. See also Peter Burke, *The French Historical Revolution: The "Annales" School, 1929–89* (Stanford, CA: Stanford University Press, 1990), 32–64.

24. See Hans-Georg Gadamer, *Truth and Method*, trans. Joel Weinsheimer and Donald G. Marshall, 2nd ed. (1975; London: Sheed and Ward, 1988), 214–18. See also Bonnie G. Smith, "Gender and the Practices of Scientific History: The Seminar and Archival Research in the Nineteenth Century," *American Historical Review* 100 (October 1995): 1150–76.

25. Braudel, "Preface to the First Edition," 20.

26. Alfred W. Crosby Jr., *The Columbian Exchange: Biological and Cultural Consequences of 1492* (1972; London: Prager, 2003), xxv.

27. See Daniel Lord Smail, *On Deep History and the Brain* (Berkeley: University of California Press, 2008), 74–189.

28. Oreskes, "Scientific Consensus," 93.

29. Alfred W. Crosby Jr., "The Past and Present of Environmental History," *American Historical Review* 100 (October 1995): 1185.

30. Jeremy Davies has recently taken me to task, rightly, for overdrawing the distinction between biological and geological agency in the original version of this chapter. See Jeremy Davies, "Noah's Dove: The Anthropocene, the Earth System and Genesis 8:8–12," *Green Letters: Studies in Ecocriticism* 23, no. 4 (2019): 337–49, https://doi.org/10.1080/14688417.2019.1706611. Indeed, I myself acknowledged as much in a later publication: "one cannot separate the biological agency of humans from their geological agency in the way in which I appeared to do in my essay, 'The Climate of History.'" Chakrabarty, "Whose Anthropocene? A Response," in *Whose Anthropocene? Revisiting Dipesh Chakrabarty's 'Four Theses,'* ed. Robert Emmett and Thomas Lekan, Rachel Carson Center Perspectives, Transformations in Environment and History, 2016/2 (Munich: Rachel Carson Center, 2016), 104. It has long been recognized that life itself works as a geological force. See Peter Westbroek, *Life as a Geological Force: Dynamics of the Earth* (New York: W. W. Norton, 1991). The globe/planet distinction that this book

turns around supersedes the distinction between biological and geological agencies I posited initially.

31. Will Steffen, director of the Centre for Resource and Environmental Studies at the Australian National University, quoted in "Humans Creating New 'Geological Age,'" *Australian*, March 31, 2008. Steffen's reference was the Millennium Ecosystem Assessment Report of 2005. See also Neil Shubin, "The Disappearance of Species," *Bulletin of the American Academy of Arts and Sciences* 61 (Spring 2008): 17–19.

32. Bill McKibben's argument about the "end of nature" implied the end of nature as "a separate realm that had always served to make us feel smaller." Bill McKibben, *The End of Nature* (1989; New York: Random House, 2006), xxii.

33. Bruno Latour's *Politics of Nature: How to Bring the Sciences into Democracy*, trans. Catherine Porter (1999; Cambridge, MA: Harvard University Press, 2004), written before the intensification of the debate on global warming, calls into question the entire tradition of organizing the idea of politics around the assumption of a separate realm of nature and points to the problems that this assumption poses for contemporary questions of democracy.

34. Gadamer, *Truth and Method*, 206. The historian "knows that everything could have been different, and every acting individual could have acted differently."

35. Paul J. Crutzen and Eugene F. Stoermer, "The Anthropocene," *IGBP Newsletter*, no. 41 (May 2000): 17.

36. Paul J. Crutzen, "Geology of Mankind," *Nature* 415, no. 23 (January 3, 2002): 23.

37. Mike Davis, "Living on the Ice Shelf: Humanity's Meltdown," TomDispatch.com, June 26, 2008, https://www.tomdispatch.com/post/174949. I am grateful to Lauren Berlant for bringing this essay to my attention.

38. Crutzen and Stoermer, "The Anthropocene," 17.

39. Crutzen and Stoermer, 17.

40. See William F. Ruddiman, "The Anthropogenic Greenhouse Era Began Thousands of Years Ago," *Climatic Change* 61, no. 3 (2003): 261–93; Paul J. Crutzen and Will Steffen, "How Long Have We Been in the Anthropocene Era?," *Climatic Change* 61, no. 3 (2003): 251–57; and Jan Zalasiewicz et al., "Are We Now Living in the Anthropocene?" *GSA Today* 18 (February 2008): 4–8.

41. Zalasiewicz et al., "Are We Now Living in the Anthropocene?," 7. Davis describes the London Society as "the world's oldest association of Earth scientists, founded in 1807"; Davis, "Living on the Ice Shelf."

42. Libby Robin and Will Steffen, "History for the Anthropocene," *History Compass* 5, no. 5 (2007): 1694–719, and Jeffrey D. Sachs, "The Anthropocene," in *Common Wealth: Economics for a Crowded Planet* (New York: Penguin, 2008), 57–82, were early instances of adoption of the term for the human sciences.

43. Edward O. Wilson, *The Future of Life* (New York: Vintage, 2002), 102.

44. Christophe Bonneuil and Jean-Baptiste Fressoz's *The Shock of the Anthropocene*, trans. David Fernbach (London: Verso, 2016) helpfully complicates the narrative here.

45. As James Hansen pointed out in his remarks at COP 25 (Madrid), "the President of Exxon Research and Engineering in 1982 correctly described the climate threat: the climate system is characterized by a delayed response and amplifying

feedbacks. Together these imply an urgency for anticipatory actions. The obvious, crucial required action was development of carbon-free energy. Instead, Exxon chose to invest in 'fracking' and continued reliance on fuels of ever greater climate footprint. They complemented this with a disinformation campaign, including a pretense that they were working hard on clean coal and renewables . . . while knowing full well that global fossil fuel emissions would continue to rise." James Hansen, "Wheels of Justice," December 26, 2019, http://www.columbia.edu/~jeh1/mailings/2019/20191226_WheelsOfJustice.pdf. See also Geoffrey Supran and Naomi Oreskes, "Assessing ExxonMobil's Climate Change Communications (1977–2014)," *Environmental Research Letters* 12 (2017), https://iopscience.iop.org/article/10.1088/1748-9326/aa816f.

46. Wilson, *The Future of Life*, 102.
47. Crutzen and Stoermer, "The Anthropocene," 18.
48. Davis, "Living on the Ice Shelf."
49. See Latour, *Politics of Nature*.
50. Wilson, *The Future of Life*, 102.
51. Flannery, *Weather Makers*, xiv.
52. Maslin, *Global Warming*, 147. For a discussion of how fossil fuels created both possibilities and limits for democracies in the twentieth century, see Timothy Mitchell, *Carbon Democracy: Political Power in the Age of Oil* (London: Verso, 2011).
53. Davis, "Living on the Ice Shelf."
54. Wilson, *In Search of Nature* (Washington, DC: Island Press, 1996), ix–x.
55. See Smail, *On Deep History and the Brain*.
56. Wilson, *In Search of Nature*, x.
57. Michael Geyer and Charles Bright, "World History in a Global Age," *American Historical Review* 100 (October 1995): 1058–59.
58. Geyer and Bright, "World History," 1059.
59. See Smail, *On Deep History and the Brain*, 124.
60. Smail, 124–25.
61. Jacques Derrida, "Cogito and the History of Madness," *Writing and Difference*, trans. Alan Bass (Chicago: University of Chicago Press, 1978), 34.
62. Sachs, *Common Wealth*, 57–82.
63. Wilson, foreword to Sachs, *Common Wealth*, xii. Students of Marx may be reminded here of the use of the category "species being" by the young Marx.
64. See Kenneth Pomeranz, *The Great Divergence: Europe, China, and the Making of the Modern World Economy* (Princeton, NJ: Princeton University Press, 2000); E. A. Wrigley, *Energy and the English Industrial Revolution* (Cambridge: Cambridge University Press, 2010); Fredrik Albritton Jonsson, "The Coal Question before Jevons," *Historical Journal* 63, no. 1 (2020): 107–26; Andreas Malm, *Fossil Capital: The Rise of Steam Power and the Roots of Global Warming* (London: Verso, 2016).
65. See Mitchell, *Carbon Democracy*. See also Edwin Black, *Internal Combustion: How Corporations and Governments Addicted the World to Oil and Derailed the Alternatives* (New York: St. Martin's, 2006).
66. Arrighi's *Long Twentieth Century* is a good guide to these fluctuations in the fortunes of capitalism.

67. Lawrence Guy Straus, "The World at the End of the Last Ice Age," in *Humans at the End of the Ice Age: The Archaeology of the Pleistocene-Holocene Transition*, ed. Lawrence Guy Straus et al. (New York: Plenum, 1996), 5.

68. Flannery, *Weather Makers*, 63, 64.

69. Since I wrote this sentence a decade ago, scientists have put forward the idea of "planetary boundaries" that humans should not breach. See Johan Rockström et al., "Planetary Boundaries: Exploring the Safe Operating Space for Humanity," *Economy and Society* 14, no. 2 (2009), http://www.ecologyandsociety.org/vol14/iss2/art32/.

70. Ashish Kothari, "The Reality of Climate Injustice," *Hindu*, November 18, 2007.

71. I have borrowed the idea of "retrospective" and "prospective" guilt from a discussion led at the Franke Institute for the Humanities by Peter Singer during the Chicago Humanities Festival, November 2007.

72. Crutzen and Stoermer, "The Anthropocene," 17.

73. See Colin Tudge, *Neanderthals, Bandits, and Farmers: How Agriculture Really Began* (New Haven, CT: Yale University Press, 1999), 35–36.

74. Wilson, *In Search of Nature*, 199.

75. Gadamer, *Truth and Method*, 232, 234. See also Michael Ermarth, *Wilhelm Dilthey: The Critique of Historical Reason* (Chicago: University of Chicago Press, 1978), 310–22.

76. See E. P. Thompson, *The Making of the English Working Class* (Harmondsworth: Penguin, 1963).

77. Louis Althusser and Étienne Balibar, *Reading Capital*, trans. Ben Brewster, 2nd ed. (London: New Left Books, 1977; first published in French, 1968), 105. This was obviously a favorite expression of Althusser's. He repeats it in his autobiography, *The Future Lasts a Long Time*, ed. Olivier Corpet and Yann Moulier Boutang, trans. Richard Vieasey (London: Vintage, 1993), 218. The original statement in Spinoza appears to occur in his *Ethics*, trans. Andrew Boyle and G. H. R. Parkinson (London: J. M. Dent & Son, 1992), First Book, Note to Proposition XVII, Corollary II, p. 19: "For the intellect and will which would constitute the essence of God must differ entirely from our will and intellect, nor can they agree in anything save name, nor any more than the dog, as a heavenly body [Sirius], and the dog, as a barking animal, agree."

78. Ursula Heise, *Imagining Extinction: The Cultural Meanings of Endangered Species* (Chicago: University of Chicago Press, 2016), 224.

79. Jacques Derrida, *The Animal That Therefore I Am*, ed. Marie-Louise Mallet and trans. David Wills (New York: Fordham University Press, 2008). The last chapter of Heise, *Imagining Extinction*, is also quite instructive on this question.

80. For a thoughtful and remarkable attempt at writing a "neo materialist" and co-evolutionary history set in the context of modern Japan and the United States, see Timothy J. LeCain, *The Matter of History: How Things Create the Past* (Cambridge: Cambridge University Press, 2017).

81. Heise, I suppose, would not disagree with this, since she writes, "there is no principled reason why it [species] cannot be translated into the realm of perception, experience, and collective self-identification by means of its own set of rhetorical, symbolic, legal, social, and institutional structures." *Imagining Extinction*, 225.

82. Geyer and Bright, "World History," 1060.

83. Michel Foucault, *The Order of Things: An Archaeology of the Human Science* (1966; New York: Vintage, 1973), 368.

84. Geyer and Bright, "World History," 1060.

85. See Alan Weisman, *The World without Us* (New York: Thomas Dunne, 2007) 25–28.

86. My use of the "lifeboat" metaphor elicited the moral ire of many critics from the Left who wanted to argue that, come what may, the rich will always have the equivalents of "lifeboats." That the rich will by definition have more resources at their disposal than the poor in any situation of crisis is something I do not deny. But the reader will notice that the point of my lifeboat metaphor in the paragraph above was to drive home the distinction between the usual boom-bust cycles that capitalism is prone to—something the rich can afford to ride out—and the climate crisis that, depending on its severity, could make even those parts of the planet that the rich love to reside in uninhabitable. My reference to fires in California and Australia was made in aid of that point. There is not much to be gained by quibbling over a literal reading of the metaphor of a lifeboat.

87. See David N. Stamos, *The Species Problem: Biological Species, Ontology, and the Metaphysics of Biology* (Lanham, MD: Lexington Books, 2003) and Jody Hey, *Genes, Categories, and Species: The Evolutionary and Cognitive Causes of the Species Problem* (New York: Oxford University Press, 2001).

88. See Theodor Adorno, "Negative Universal History," in *History and Freedom: Lectures 1964–1965*, ed. Rolf Tiedemann, trans. Rodney Livingstone (2006; Cambridge: Polity, 2009; originally in German, 2001). Antonio Y. Vasquez-Arroyo's article "Universal History Disavowed: On Critical Theory and Postcolonialism," *Postcolonial Studies* 11, no. 4 (December 2008), 451–73, originally inspired this proposition. See also the lucid discussion in Harriet Johnson, "The Anthropocene as a Negative Universal History," *Adorno Studies* 3, no. 1 (July 2019): 47–63.

89. Heise, *Imagining Extinction*, 224.

90. Jason Moore, *Capitalism in the Web of Life* (London: Verso, 2015), ix.

91. John Bellamy Foster, Brett Clark, and Richard York, *The Ecological Rift: Capitalism's War on the Earth* (New York: Monthly Review Press, 2010), 395.

92. Adorno, "Negative Universal History," 96.

93. Johnson, "Anthropocene as a Negative Universal History," 60–61.

94. This is indeed the problem that Latour addresses in much of his work.

95. On all this, see Philippe Descola, *Beyond Nature and Culture*, trans. Janet Lloyd (Chicago: University of Chicago Press, 2013; first published in French, 2005); Eduardo Vivieros de Castro, *Cannibal Metaphysics: For a Post-Structural Anthropology*, ed. and trans. Peter Skafish (Minneapolis, MN: Univocal, 2014; first published in French, 2009); Deborah Bird Rose, *Dingo Makes Us Human: Life and Land in an Australian Aboriginal Culture* (Cambridge: Cambridge University Press, 2000).

96. See the last chapter of this book.

CHAPTER TWO

1. Bryan Lovell, *Challenged by Carbon: The Oil Industry and Climate Change* (New York: Cambridge University Press, 2010), 75.

2. See Intergovernmental Panel on Climate Change, *Climate Change 2007: The Physical Science Basis*, ed. Susan Solomon et al. (2007; Cambridge: Cambridge University Press, 2008), 446, box 6.2.

3. Lovell, *Challenged by Carbon*, xi.

4. David Archer, *The Long Thaw: How Humans Are Changing the Next 100,000 Years of Earth's Climate* (Princeton, NJ: Princeton University Press, 2009), 6.

5. Archer, 11.

6. See Curt Stager, *Deep Future: The Next 100,000 Years of Life on Earth* (New York: St. Martin's, 2011), chap. 2.

7. David Archer, *The Global Carbon Cycle* (Princeton, NJ: Princeton University Press, 2010), 21.

8. A thoughtful series of essays connecting public perceptions of risks with their management through statistical analyses and political and legal regulation is to be had in Cass R. Sunstein, *Risk and Reason: Safety, Law, and the Environment* (New York: Cambridge University Pres, 2002).

9. Charles S. Pearson, *Economics and the Challenge of Global Warming* (New York: Cambridge University Press, 2011), 25n6.

10. A classic text on this topic is Frank H. Knight, *Risk, Uncertainty, and Profit* (1921; London: Forgotten Books, 2012). Knight would have objected to my use of the word *art* with regard to the discipline of economics, for he considered it to be part of the sciences. He begins the book with the statement "Economics, or more properly theoretical economics, is the only one of the social sciences which has aspired to the distinction of an exact science" while praising physics for securing "our present marvelous mastery over the forces of nature" (3, 5). On the whole question of the role of uncertainty in various aspects of modern life, see Helga Nowotny, *The Cunning of Uncertainty* (Cambridge: Polity, 2016).

11. See, for example, the chart reproduced in Nicholas Stern and Michael Jacobs, *The Economics of Climate Change: The Stern Review* (New York: Cambridge University Press, 2007), 200. See also Eric A. Posner and David Weisbach, *Climate Change Justice* (Princeton, NJ: Princeton University Press, 2010), chap. 2.

12. In a series of essays, the late Martin Weitzman emphasized how the usual cost-benefit analyses of *welfare loss* due to climate change assume temperature rises on the lower side; the uncertainties of calculating the *damage function* consequent on a catastrophic rise of 10°–20°C in the average global surface temperature throw economic calculations haywire. Weitzman remarks: "Even just acknowledging more openly the incredible magnitude of the deep structural uncertainties . . . involved in climate-change analysis—and explaining better to policy makers that the artificial crispness conveyed by conventional IAM [Integrated Assessment Model]-based CBAs [cost-benefit analyses] . . . is especially and unusually misleading compared with more-ordinary non-climate-change CBA situations—might elevate the level of public discourse concerning what to do about global warming." Martin L. Weitzman, "Some Basic Economics of Extreme Climate Change," Harvard Environmental Economics Program, Discussion Paper 09-10, February 19, 2009, p. 26, https://scholar.harvard.edu/files/weitzman/files/heep_discussion_10.pdf. See also Weitzman, "GHG Targets as Insurance against Catastrophic Climate Damages," *Journal of Public Economic Theory* 14 (March 2012): 221–44.

13. Archer, *Global Carbon Cycle*, 22. Lovelock himself defends the concept of Gaia at least as a metaphor; see James Lovelock, *The Vanishing Face of Gaia* (New York: Basic Books, 2009), 13. The recent collaborative work of Bruno Latour and Tim Lenton on

Gaia expands on and develops this vitalist aspect of the problem. This is why they make a critical distinction between Gaia and Earth System Science. See Bruno Latour and Timothy N. Lenton, "Extending the Domain of Freedom, or Why Gaia Is So Difficult to Understand," *Critical Inquiry* 45, no. 3 (Spring 2019): 659–80. See also Timothy Lenton, Sébastien Dutreuil, and Bruno Latour, "Life on Earth Is Hard to Spot," *Anthropocene Review* (forthcoming).

14. Archer, *Global Carbon Cycle*, 1.

15. Wallace S. Broecker and Robert Kunzig, *Fixing Climate: What Past Climate Changes Reveal about the Current Threat—and How to Counter It* (New York: Hill and Wang, 2008), 100.

16. James Hansen, *Storms of My Grandchildren: The Truth about the Coming Climate Catastrophe and Our Last Chance to Save Humanity* (New York: Bloomsbury, 2009), 71.

17. Archer, *Long Thaw*, 95.

18. John Broome, *Climate Matters: Ethics in a Warming World* (New York: W. W. Norton, 2012), 128, 129.

19. Paul N. Edwards, *A Vast Machine: Computer Models, Climate Data, and the Politics of Global Warming* (Cambridge, MA: MIT Press, 2010), 438–39, emphasis added.

20. Edwards, *Vast Machine*, 431.

21. Edwards, 439.

22. Pearson, *Economics*, 26, 31.

23. Pearson, 30.

24. Sunstein, *Risk and Reason*, 103.

25. Hansen, *Storms of My Grandchildren*, 176.

26. Sunstein, *Risk and Reason*, 129n40.

27. See Pearson, *Economics*. Sunstein acknowledges that "the worst-case scenario involving global warming" calls for the application of the maximin principle and yet recommends the "'cap-and-trade' system"—which assumes a gradual transition to renewables—as it "seems to be the most promising, in part because it is so much less expensive than the alternatives." Sunstein, *Risk and Reason*, 129. This amounts to replacing the maximin principle by the precautionary one. We can only infer how little understood the challenge of global-warming-related "uncertainty" was among scholars who assumed that the usual strategies of risk management would be an adequate response to the problem.

28. See Stephen M. Gardiner, "Cost-Benefit Paralysis," *A Perfect Moral Storm: The Ethical Tragedy of Climate Change* (New York, 2011), chap. 8.

29. Peter Newell and Matthew Paterson, *Climate Capitalism: Global Warming and the Transformation of the Global Economy* (New York: Oxford University Press, 2010), 7, emphasis added.

30. Sunita Narain, blurb for Newell and Paterson, *Climate Capitalism*, back cover.

31. John Bellamy Foster, Brett Clark, and Richard York, *The Ecological Rift: Capitalism's War on the Earth* (New York: Monthly Review Press, 2010), 47.

32. See Sha Zukang, "Overview," in United Nations, Department of Economic and Social Affairs, *Promoting Development, Saving the Planet* (New York: United Nations Publishing Section, 2009), https://www.un.org/en/development/desa/policy/wess/wess_archive/2009wess.pdf.

33. Sha, viii.

34. Sha, xviii.

35. Sha, 3.

36. See Anil Agarwal and Sunita Narain, *Global Warming in an Unequal World: A Case of Environmental Colonialism* (1991; repr., New Delhi: Centre for Science and Environment, 2003).

37. "Report of the United Nations Conference on Environment and Development (Rio de Janeiro, 3–14 June 1992), Annex I, Rio Declaration on Environment and Development." Principle 7, August 12, 1992, https://www.un.org/en/development/desa/population/migration/generalassembly/docs/globalcompact/A_CONF.151_26_Vol.I_Declaration.pdf. Kyoto Protocol to the United Nations Framework Convention on Climate Change, Article 10, December 10, 1997, https://unfccc.int/sites/default/files/resource/docs/cop3/l07a01.pdf.

38. Agarwal and Narain, *Global Warming in an Unequal World*, 4–9, quote 9.

39. A generative history of the "global population" question is provided in Alison Bashford, *Global Population: History, Geopolitics, and Life on Earth* (New York: Columbia University Press, 2014).

40. Jairam Ramesh et al., "Climate Change and Parliament," in *Handbook of Climate Change and India: Development, Politics, and Governance*, ed. Navroz K. Dubash (New York: Earthscan, 2012), 238. D. Raghunandan argues that this "climate justice" position that India championed at many international forums on climate change was informed more by "geopolitical assessments" than by any "deep scientific understanding." D. Raghunandan, "India's Official Position: A Critical View Based on Science," in Dubash, *Handbook of Climate Change and India*, 172, 173.

41. Quoted in Y. P. Anand and Mark Lindley, "Gandhi on Providence and Greed," 1, https://www.academia.edu/303042/Gandhi_on_providence_and_greed. Gandhi is supposed to have said this in Hindi in 1947 to his secretary, Pyarelal Nayyar, who reproduced it in his book *Mahatma Gandhi: The Last Phase* (Ahmedabad: Navajivan, 1958), 2:552. Anand and Lindley say that Gandhi was influenced by the work of J. C. Kumarappa, in turn a Gandhian economist to whose book *Economy of Permanence* (1945) Gandhi contributed a preface. Interestingly, India's *National Action Plan on Climate Change* incorrectly paraphrases Gandhi's dictum as saying "the earth has enough resources to meet people's needs, but will never have enough to satisfy people's greed," thus missing the emphasis that Gandhi typically put on the individual's sense of moral responsibility. Government of India, *National Action Plan on Climate Change*, 1, http://moef.gov.in/wp-content/uploads/2018/04/NAP_E.pdf.

42. Peter Galuszka, "With China and India Ravenous for Energy, Coal's Future Seems Assured," *New York Times*, November 12, 2012, https://www.nytimes.com/2012/11/13/business/energy-environment/china-leads-the-way-as-demand-for-coal-surges-worldwide.html?_r=0.

43. Galuszka.

44. Amitav Ghosh, *The Great Derangement: Climate Change and the Unthinkable* (Chicago: University of Chicago Press, 2016), pt. 3 on "Politics."

45. P. W. Anderson, "More Is Different: Broken Symmetry and the Nature of the Hierarchical Structure of Science," *Science*. 177, no. 4047 (August 4, 1972): 393–96. Thanks to Sabyasachi Bhattacharya for directing me to this enormously insightful paper and for discussing its implications.

46. See Vaclav Smil, *Harvesting the Biosphere: What We Have Taken from Nature* (Cambridge, MA: MIT Press, 2013), 221, and Tom Butler, Daniel Lerch, and George Wuerthner, "Introduction: Energy Literacy," in *The Energy Reader: Overdevelopment and the Delusion of Endless Growth*, ed. Thomas Butler, Daniel Lerch, and George Wuerthner (Sausalito, CA: Watershed Media, 2012), 11–12.

47. Some of the complexities of this problem are captured nicely in the chapter "Outsider Monkey, Insider Monkey" in Radhika Govindrajan, *Animal Intimacies: Interspecies Relatedness in India's Central Himalayas* (Chicago: The University of Chicago Press, 2018), chap. 4. I am grateful to Sneha Annavarapu for drawing my attention to this pioneering study.

48. Hansen, *Storms of My Grandchildren*.

49. Curt Stager, *Deep Future: The Next 100,000 Years of Life on Earth* (New York: St. Martin's, 2011), 62–66. See also the discussion in Hansen, *Storms of My Grandchildren*, 145–46.

50. See Michael Denny and Lisa Matisoo-Smith, "Rethinking Polynesian Origins: Human Settlement of the Pacific," LENScience, Senior Biology Seminar Series, http://www.hokulea.com/wp-content/uploads/2015/03/Rethinking-Polynesian-origins.pdf.

51. Burton Richter, *Beyond Smoke and Mirrors: Climate Change and Energy in the Twenty-First Century* (New York: Cambridge University Press, 2010), 2. The long story of human adaptation to a variety of climates since the advent of *Homo sapiens* is masterfully told in John L. Brooks, *Climate Change and the Course of Global History: A Rough Journey* (Cambridge: Cambridge University Press, 2014).

52. See P. K. Haff, "Technology as a Geological Phenomenon: Implications for Human Well-Being," in "A Stratigraphical Basis for the Anthropocene," ed. C. N. Waters et al., special issue, *Geological Society of London* 395 (2014): 301–9, https://sp.lyellcollection.org/content/395/1/301.

53. See Will Steffen, Paul J. Crutzen, and John R. McNeill, "The Anthropocene: Are Humans Now Overwhelming the Great Forces of Nature?" *Ambio* 36 (December 2007): 614–21.

54. For an Australian example of this, see Lesley Johnson, *The Modern Girl: Girlhood and Growing Up* (St. Leonards, New South Wales: Allen and Unwin, 1993).

55. See Assa Doron and Robin Jeffrey, *The Great Indian Phone Book: How the Cheap Cell Phone Changes Business, Politics, and Daily Life* (Cambridge, MA: Harvard University Press, 2013).

56. Jan Zalasiewicz, "The Human Touch," *Paleontology Newsletter* 82 (March 2013): 24, https://www.palass.org/sites/default/files/media/publications/newsletters/number_82/number82.pdf. While Zalasiewicz's summary of Smil's researches is extremely helpful, it should be remembered that most of Smil's effort is directed at reminding the reader of the methodological challenges involved in measuring the changes reported here and how approximate and provisional the relevant numbers are. Zalasiewicz's figures are based on Smil, "Harvesting the Biosphere: The Human Impact," *Population and Development Review* 37 (December 2011): 613–36.

57. Smil, "Harvesting the Biosphere," p. 252.

58. John Broome, *Climate Matters: Ethics in a Warming World* (New York: W. W. Norton, 2012), 112–13.

59. Archer, *Long Thaw*, 2.

60. Émile Durkheim, *The Elementary Forms of Religious Life*, trans. Joseph Ward Swain (1915; Mineola, NY: Dover, 2008), 134, 139.

61. Fazlur Rahman, *Major Themes of the Qur'an* (Chicago: University of Chicago Press, 2009), 12–13.

62. Rahman, 13.

63. An interesting text claiming—from a mixture of Hindu and Buddhist perspectives—a special relationship between man and God is Rabindranath Tagore's 1930 Oxford Hibbert Lectures, published as "The Religion of Man" (1931), in which Tagore showed an awareness of a Hindu theological position that conceived of God as indifferent to human affairs but rejected it in favor of a Buddhist understanding of infinity that "was not the idea of a spirit of an unbounded cosmic activity, but the infinite whose meaning is in the positive ideal of goodness and love, which cannot be otherwise than human." Rabindranath Tagore, "The Religion of Man," in *The English Writings of Rabindranath Tagore*, vol. 3, *Miscellany*, ed. Sisir Kumar Das (New Delhi: Sahitya Akademi, 1999), 111.

64. Ernest Partridge, "Nature as a Moral Resource," *Environmental Ethics* 6 (Summer 1984): 103.

65. See, for instance, Lawrence Buell, "The Misery of Beasts and Humans: Nonanthropocentric Ethics versus Environmental Justice," *Writing for an Endangered World: Literature, Culture, and Environment in the U.S. and Beyond* (Cambridge, MA: Belknap Press of Harvard University Press, 2001), 224–42.

66. Feng Han, "The Chinese View of Nature: Tourism in China's Scenic and Historic Interest Areas" (PhD diss., Queensland University of Technology, 2008), 22–23, https://eprints.qut.edu.au/16480/1/Feng_Han_Thesis.pdf. I am grateful to Ken Taylor for drawing my attention to this thesis. Han, of course, is echoing Eugene Hargrove; see Eugene C. Hargrove, "Weak Anthropocentric Intrinsic Value," *Monist* 75 (April 1992): 183–207, and Karyn Lai, "Environmental Concern: Can Humans Avoid Being Partial? Epistemological Awareness in the Zhuangzi," in *Nature, Environment, and Culture in East Asia: The Challenge of Climate Change*, ed. Carmen Meinert (Leiden: Brill, 2013), 79.

67. See, for example, Bryan G. Norton, "Environmental Ethics and Weak Anthropocentrism," *Environmental Ethics* 6 (Summer 1984): 131–48. Norton was the first to propose the idea of weak anthropocentrism that has since been taken up by many.

68. James Lovelock, *The Vanishing Face of Gaia* (New York: Basic Books, 2009), 35–36.

69. Lovelock, 36.

70. Slavoj Žižek, *Living in the End Times* (New York: Verso, 2010), 332, 333–34. See also chapter 1 above.

71. Ursula Heise, *Imagining Extinction: The Cultural Meanings of Endangered Species* (Chicago: University of Chicago Press, 2016), 223.

72. See Stager, *Deep Future*, chap. 2.

73. Gayatri Chakravorty Spivak, *An Aesthetic Education in the Era of Globalization* (Cambridge, MA: Harvard University Press, 2012), 338.

74. See Raymond T. Pierrehumbert, *Principles of Planetary Climate* (New York University Press: Cambridge, 2010).

75. I speak of the growing divergence between the planetary and the global because there is an established tradition of using the two words to mean the same thing. See, for instance, Carl Schmitt, *The Nomos of the Earth in the International Law of the Jus Publicum Europaeum*, trans. G. L. Ulmen (New York: Telos Press, 2006), 86–88, 173, 351, and chapter 3 below.

CHAPTER THREE

1. The phrase "regimes of historicity" registers my debt to François Hartog, from whose work I borrow the idea. The word *regime* implies some kind of ordering, the ordering of historical time. "Why 'regime' rather than 'form'?" asks Hartog. He answers the question by referring to the word's association in French with "the idea of degrees . . . of mixtures and composites, and an always provisional or unstable equilibrium" and thus to a provisional state of order. François Hartog, *Regimes of Historicity: Presentism and Experiences of Time*, trans. Saskia Brown (New York: Columbia University Press, 2015), xv.

2. Martin Heidegger, "The Origin of the Work of Art," in *Poetry, Language, Thought*, trans. Albert Hofstadter (New York: Harper and Row, 1975), 42. I should clarify that I use the capitalized form of the word *Earth* (unless it is part of a quotation) to designate an abstract and *unvisualizable* entity put together by practitioners of Earth System Science. In all other cases, including the idea of the earth as a globe—the "blue marble" picture, say, of 1972—I use the lowercase word *earth*.

3. Heidegger, 46.

4. Heidegger, "Building Dwelling Thinking," in *Poetry, Language, Thought*, 149.

5. Heidegger, 150.

6. Heidegger, "Origin of the Work of Art," 48–49. I should make it clear that my employment of Heideggerian terms like *earth* or *world* is conceptual and not philological. In other words, I assume that our capacity to understand Heidegger's concepts is never fatally crippled by the fact that not all languages may possess words that correspond exactly to those that Heidegger deployed.

7. Hans-Georg Gadamer, "The Truth of the Work of Art," in *Heidegger's Ways*, trans. John W. Stanley (Albany, NY: State University of New York Press, 1994), 99.

8. Gadamer, "Martin Heidegger—85 Years," in Gadamer, *Heidegger's Ways*, 117.

9. Andrew S. Goudie and Heather A. Viles, *Geomorphology in the Anthropocene* (New York: Cambridge University Press, 2016), p. 7.

10. Heidegger, "Origin of the Work of Art," 47.

11. Heidegger writes, "We come and stand facing a tree, before it, and the tree faces, meets us. Which one is meeting here? The tree, or we? Or both? Or neither? We come and stand—just as we are, and not merely with our head or our consciousness—facing the tree in bloom, and the tree faces, meets us as the tree it is." Heidegger, *What Is Called Thinking?* trans. J. Glenn Gray (1954; New York: Harper and Row, 2004), 42.

12. Heidegger posits a dyadic relationship of mutuality between the world and humans (only after the latter has come into language; presumably humans without language were the same as other animals) whereby language in Heidegger's expression becomes "the house of Being" and humans those to whom the coming to presence of Being is vouchsafed. Thus, the essence of Being—that is, the question of Being—becomes tied to the human, "because his essence is to be one who waits, the one who

attends upon the coming to presence of Being in that in thinking he guards it." Martin Heidegger, "The Turning," in *The Question Concerning Technology and Other Essays*, trans. William Lovitt (New York: Harper Torchbooks, 1977), 42. The crisis of this assumed relationship of mutuality is what I discuss in more detail in chapter 8.

13. Poetically, and politically, Latour gives Gaia a persona and a face to enable humans to face Gaia; see Bruno Latour, *Facing Gaia: Eight Lectures on the New Climatic Regime*, trans. Catherine Porter (Medford, MA: Polity, 2017), 280–84.

14. Kant's 1756 essays on earthquakes are fascinating in this regard. Immanuel Kant, "On the Causes of Earthquakes on the Occasion of the Calamity That Befell the Western Countries of Europe towards the End of Last Year," "History and Natural Description of the Most Noteworthy Occurrences of the Earthquake That Struck a Large Part of the Earth at the End of the Year 1755," and "Continued Observations on the Earthquakes That Have Been Experienced for Some Time" (1756), all translated by Olaf Reinhardt in Immanuel Kant, *Natural Science*, ed. Eric Watkins (Cambridge: Cambridge University Press, 2012), 330–36, 339–64, 367–73. See also Edgar S. Brightman, "The Lisbon Earthquake: A Study in Religious Valuation," *American Journal of Theology* 23 (October 1919), 500–518; José Oscar de Almeida Marques, "The Paths of Providence: Voltaire and Rousseau on the Lisbon Earthquake," *Cadernos de Cadernos de História e Filosofia da Ciência* 15, no. 1 (2005): 33–57; and Dipesh Chakrabarty, "The Power of Superstition in Public Life in India," *Economic and Political Weekly*, May 17, 2008, 16–19.

15. See Andrea Westermann, "Disciplining the Earth: Earthquake Observation in Switzerland and Germany at the Turn of the Nineteenth Century," and Frank Oberholzner, "From an Act of God to an Insurable Risk: The Change in the Perception of Hailstorms and Thunderstorms since the Early Modern Period," *Environment and History* 17 (February 2011): 53–77, 133–52.

16. See Carl Schmitt, *Dialogues on Power and Space*, trans. Samuel Garrett Zeitlin, ed. Frederico Finchelstein and Andreas Kalyvas (Malden, MA: Polity, 2015). Thanks to Bruno Latour for drawing my attention to this text.

17. Spivak's thoughts about planetarity are elaborated on in her *Death of A Discipline* (New York: Columbia University Press, 2003). For more on Spivak's insights into planetarity, see Elizabeth M. DeLoughrey, "Planetary Militarized Radiations," *Allegories of the Anthropocene* (Durham, NC: Duke University Press, 2019), 63–97; Benjamin Morgan, "*Fin du Globe*: On Decadent Planets," *Victorian Studies* 58, no. 4 (Summer 2016): 609–35. See also Eugene Thacker's *In the Dust of This Planet* (Washington, DC: Zero Books, 2011) for some other stimulating ideas about planetarity.

18. The following section expands and elaborates on a proposition I first put forward in "Planetary Crises and the Difficulty of Being Modern," *Millennium: Journal of International Studies* 46, no. 3 (2018): 259–82.

19. I am indebted to Catherine Malabou for the articulation of this formulation. See Chakrabarty, "Afterword," *South Atlantic Quarterly* 116 (January 2017): 166.

20. See Joyce E. Chaplin, *Round about the Earth: Circumnavigation from Magellan to Orbit* (New York: Simon and Schuster, 2013).

21. Thomas Hobbes, *Leviathan*, ed. and with an introduction by C. B. Macpherson (1651; Harmondsworth: Penguin, 1976), 186.

22. Hannah Arendt, *The Human Condition*, 2nd ed., with an introduction by Margaret Canovan (1958; Chicago: University of Chicago Press, 1998), 250.

23. See Carl Schmitt, *The Nomos of the Earth: In the International Law of the Jus Publicum Europaeum*, trans. G. L. Ulman (New York: Telos, 2003).

24. Also see the discussion in Schmitt, *Dialogues on Power and Space*.

25. Schmitt, *Nomos of the Earth*, 328.

26. Schmitt, 329.

27. Schmitt, 352.

28. Schmitt, 87.

29. Schmitt, 88.

30. Schmitt, 173.

31. Schmitt, 351.

32. Schmitt, *Land and Sea*, trans. Simona Draghici (1954; Washington, DC: Plutarch Press, 1997), 11.

33. See Benjamin Lazier, "Earthrise; or, The Globalization of the World Picture," *American Historical Review* 116 (June 2011): 602–30, and Kelly Oliver, "The Earth's Refusal: Heidegger," in *Earth and World: Philosophy after the Apollo Missions* (New York: Columbia University Press, 2015), 111–62.

34. Heidegger, "The Age of the World Picture," in *The Question Concerning Technology and Other Essays*, trans. William Lovitt (New York: Harper and Row, 1977), 152.

35. Heidegger, "Plato's Doctrine of Truth," trans. Thomas Sheehan, in *Pathmarks*, trans. Sheehan et al., ed. William McNeill (1967; New York: Cambridge University Press, 1998), 182.

36. Heidegger, "Age of the World Picture," 129–30. See appendix 10 (p. 153) to this essay for Heidegger's gloss on "anthropology." For more on discussion of Heidegger's use of the words *earth*, *world*, and *planet*, see Oliver, "Earth's Refusal." See also Dana R. Villa, "The Critique of Modernity," *Arendt and Heidegger: The Fate of the Political* (Princeton, NJ: Princeton University Press, 1996), 171–208.

37. See J. R. McNeill and Peter Engelke, *The Great Acceleration: An Environmental History of the Anthropocene since 1945* (Cambridge, MA: Belknap Press of Harvard University Press, 2014).

38. See Raymond T. Pierrehumbert, *Principles of Planetary Climate* (New York: Cambridge University Press, 2010). Thus, as geologist colleagues point out, there exist university departments devoted to studying "Earth and Planetary Sciences" to include work on other planets done following earth science methods (not those of astronomy).

39. James Hansen, *Storms of My Grandchildren: The Truth about the Coming Climate Catastrophe and Our Last Chance to Save Humanity* (New York: Bloomsbury, 2009), xiv–xv.

40. See Spencer R. Weart, *The Discovery of Global Warming* (Cambridge, MA: Harvard University Press, 2008); Joshua P. Howe, *Behind the Curve: Science and the Politics of Global Warming* (Seattle: University of Washington Press, 2014); Clive Hamilton, *Defiant Earth: The Fate of Humans in the Anthropocene* (Malden, MA: Polity, 2017); and Ian Angus, *Facing the Anthropocene: Fossil Capitalism and the Crisis of the Earth System* (New York: Monthly Review Press, 2016). See also Joseph Masco, "Bad Weather: On Planetary Crisis, " *Social Studies of Science* 40 (February 2020): 7–40; DeLoughrey, "Planetarity Militarized Radiations"; and Perrin Selcer, *The Postwar Origins of the Global Environment: How the United Nations Built Spaceship Earth* (New York: Columbia University Press, 2018).

41. See Weart, *Discovery of Global Warming*, 144–45.

42. Tim Lenton, *Earth System Science: A Very Short Introduction* (New York: Oxford University Press, 2016), 1.

43. The International Geosphere-Biosphere Programme, "Earth System Definitions," http://www.igbp.net/globalchange/earthsystemdefinitions.4.d8b4c3c12bf3be638a80001040.html.

44. Quoted in Angus, *Facing the Anthropocene*, 29.

45. See Bruno Latour's fascinating discussion of this problem in Latour, "Third Lecture: Gaia, a (Finally Secular) Figure for Nature," in Latour, *Facing Gaia*, 75–110. See also Bruno Latour and Tim Lenton, "Extending the Domain of Freedom, or Why Gaia Is So Hard to Understand," *Critical Inquiry* 45 (Spring 2019): 659–80.

46. Timothy M. Lenton and Andrew Watson, *Revolutions That Made the Earth* (New York: Oxford University Press, 2011), vii.

47. Lenton, *Earth System Science*, 17. "We should recognize that Gaia is not a globe at all but a thin biofilm, a surface, a pellicle no more than a few kilometers thick that has not made inroads very far up in the atmosphere nor very far down in the deep earth below, no matter how long you consider the history of life forms. That is why it is important to shift from the global vision of Gaia to what some scientists now call the 'critical zone.'" Latour and Lenton, "Extending the Domain of Freedom," 676.

48. Lee R. Kump, James F. Kasting, and Robert G. Crane, *The Earth System* (Upper Saddle River, NJ: Prentice Hall, 2004), 3, xi, emphasis added. Jan Zalasiewicz writes, "It is true that the Earth surface is where the most immediate and significant (to us, now) processes take place, but most of the fundamental chemical cycles include shorter and longer detours and modifications within the Earth's surface, certainly down to the deeper mantle in some instances and perhaps further. Most of the Earth's water may have been derived from the Earth's mantle (and most of our oceans seem to be slowly being subducted back there, albeit very slowly, on a billion-year timescale). Shallower zones within the crust/lithosphere are active on shorter, though still geological, timescales." Jan Zalasiewicz, email to author, October 6, 2018.

49. See Zalasiewicz, email to author, October 6, 2018. Zalasiewicz thinks that the deeper parts of the world are definitely a part of the Earth system.

50. Quoted in Erle C. Ellis, *Anthropocene: A Very Short Introduction* (New York: Oxford University Press, 2018), 31, 32.

51. Lovelock himself writes, "The idea of an Earth system science . . . came into my mind at the Jet Propulsion Laboratory in California in September 1965. The first paper to mention it was published in the *Proceedings of the American Astronautical Society* in 1968. . . . The Gaia hypothesis arose in the period before it received its name." James Lovelock, *The Vanishing Face of Gaia: A Final Warning* (New York: Basic Books, 2009), 159. But he considered the name of this science "anodyne," for, while he regarded the relationship between Earth System Science and the Gaia theory as "friendly," "to understand Gaia," he thought, "requires an *instinctive* familiarity with the dynamics of systems in action, and this not a normal part of Earth or life science" 161, 167, emphasis added). See also Lovelock, "What Is Gaia?," in *The Revenge of Gaia: Earth's Climate Crisis and the Fate of Humanity* (New York: Basic Books, 2007), 15–38.

52. Lovelock, *Gaia: A New Look at Life on Earth* (1979; New York: Oxford University Press, 1995), 7–8, emphasis added.

53. Ayesha Ramachandran, *The Worldmakers: Global Imagining in Early Modern Europe* (Chicago: University of Chicago Press, 2015), 24.

54. Ramachandran, 56. Thanks also to David Orsbon, who kindly let me read his unpublished "The Person of Natura" (2017). Sverre Raffnsøe observes in an email communication, however, that while "the original Stoic conception of *kataskopos* can certainly be described as a 'view from above,'" he does not think that "it can be characterized as a view 'from without' already in antiquity. In Cicero's *Somnium Scipionis*, Scipio Aemilianus still 'only' finds himself looking down from the highest place in the world to find Carthage and Rome dwarfed. Only later can the Christianized viewer truly aim to become a 'spectator from without.'" Sverre Raffnsøe, email to author, July 9, 2019.

55. See the discussion in Ronald Weber, *Seeing Earth: Literary Responses to Space Exploration* (Athens: Ohio University Press, 1985).

56. Lenton and Watson, *Revolutions*, 301.

57. Lovelock, *Gaia*, 8.

58. See Timothy Morton, *Hyperobjects: Philosophy and Ecology after the End of the World* (Minneapolis: University of Minnesota Press, 2013).

59. Delf Rothe, "Global Security in A Posthuman Age? IR and the Anthropocene Challenge," in *Reflections on the Posthuman in International Relations: The Anthropocene, Security and Ecology*, ed. Clara Eroukhmanoff and Matt Harker (Bristol: E-International Relations, 2017), 92. Timothy Morton's recent gloss on the Heideggerian word *withdrawn* is helpful here: "'Withdrawn' doesn't mean empirically shrunken back or moving behind; it means . . . *so in your face that you can't see it*." Timothy Morton, *Humankind: Solidarity with Nonhuman People* (London: Verso, 2019), 37, emphasis in original.

60. Lenton and Watson, *Revolutions*, vii–viii. See Latour and Lenton, "Extending the Domain of Freedom," and my conversation with Latour reproduced at the end of the book.

61. Hannah Arendt, *The Human Condition*, 2nd ed., with an introduction by Margaret Canovan (1958; Chicago: University of Chicago Press, 1998), 264.

62. See Lazier, "Earthrise," and Chakrabarty, "The Human Condition in the Anthropocene," in vol. 35 of *The Tanner Lectures on Human Values*, ed. Mark Matheson (Salt Lake City: University of Utah Press, 2016), 137–88.

63. Lovelock, *Gaia*, xiv.

64. John Milton, *Paradise Lost*, ed. John Leonard (New York: Penguin Books, 2000), 20.

65. See Mary R. Albert and Geoffrey Hargreaves, "Drilling through Ice and into the Past," *Oilfield Review* 25 (Winter 2013/2014): 4–15; P. G. Talalay, "Perspectives for Development of Ice-Core Drilling Technology: A Discussion," *Annals of Glaciology* 55, no. 68 (2014): 339–50; and Richard B. Alley, "Going to Greenland," *The Two-Mile Time Machine: Ice Cores, Abrupt Climate Change, and Our Future* (Princeton, NJ: Princeton University Press, 2000), 17–30.

66. See the discussion in Paul Warde, "The Invention of Sustainability," *Modern Intellectual History* 8, no. 1 (2011): 153–70. Warde has since spelled out his larger and fascinating argument in *The Invention of Sustainability: Nature and Destiny, c. 1500–1870* (New York: Cambridge University Press, 2018).

67. Quoted in Stephen Morse, *Sustainability: A Biological Perspective* (New York: Cambridge University Press, 2010), 6. Emma Rothschild, "Maintaining (Environmental) Capital Intact," *Modern Intellectual History* 8, no. 1 (2011): 193–212, draws interesting connections between modern economists' discussion of sustainability and their debates on capital theory in the 1920s and 1930s. Deanna K. Kreisel writes, citing the *Oxford English Dictionary*, that "the term 'sustainable' was not used in the sense of minimizing environmental impact until 1976, and was not used to mean 'capable of being maintained at a certain level' until 1924." Deanna K. Kreisel, "'Form against Force': Sustainability and Organicism in the Work of John Ruskin," in *Ecological Form: System and Aesthetics in the Age of Empire*, ed. Nathan K. Hensley and Philip Steer (New York: Fordham University Press, 2019), 105.

68. Warde, "The Invention of Sustainability," 153.

69. Warde, 168, 170. Karl Marx's deep interest in von Liebig's work is noted in Paul Burkett, "Introduction to the Haymarket Edition," *Marx and Nature: A Red and Green Perspective* (Chicago: Haymarket Books, 2014), xix, and discussed in detail in John Bellamy Foster, *Marx's Ecology: Materialism and Nature* (New York: Monthly Review Press, 2000).

70. Donald Worster, *Shrinking the Earth: The Rise and Decline of Natural Abundance* (New York: Oxford University Press, 2016), 140, 141, emphasis added. For more on Vogt, see Selcer, *Postwar Origins of the Global Environment*, 68–70.

71. Quoted in Worster, 140. For intellectual and institutional histories of the idea of *environment*, see Paul Warde, Libby Robin, and Sverker Sörlin, *The Environment: A History of the Idea* (Baltimore, MD: Johns Hopkins University Press, 2018), and Selcer, *Postwar Origins of the Global Environment*.

72. For a sustained critique of the neoliberal adoption of the idea or slogan of sustainability, see Ruth Irwin, *Heidegger, Politics and Climate Change: Risking It All* (New York: Continuum, 2008).

73. P. A. Larkin, "An Epitaph for the Concept of Maximum Sustainable Yield," *Transactions of the American Fisheries Society* 106 (January 1977): 1–2. An excellent article documenting the overly political and economic nature of biology as applied to fisheries management in Europe and North America is Jennifer Hubbard's "In the Wake of Politics: The Political and Economic Construction of Fisheries Biology, 1860–1970," *Isis* 105 (June 2014): 364–78. For a brief biographical note on Larkin, see the entry in Social Networks and Archival Context, https://snaccooperative.org/ark:/99166/w6fj6xxx.

74. Morse, *Sustainability*, 5–6.

75. Jeffrey T. Nealon, *Plant Theory: Biopower and Vegetable Life* (Stanford, CA: Stanford University Press, 2016), 53–54.

76. Charles H. Langmuir and Broecker, *How to Build a Habitable Planet: The Story of Earth from the Big Bang to Humankind* (Princeton, NJ: Princeton University Press, 2012), 537.

77. "It has been modeled to go above this level—perhaps to some 30 percent in the Carboniferous—or below it (in the putative 'oxygen crisis' of the Permian-Triassic boundary)." Zalasiewicz, email to author, October 6, 2018.

78. Langmuir and Broecker, *How to Build a Habitable Planet*, 458.

79. Kump, Kasting, and Crane, *Earth System*, 225.

80. Lenton, *Earth System Science*, 44.

81. See Lenton, *Earth System Science*, 44–46; Lovelock, *Gaia* 6, 59–77; Langmuir and Broecker, *How to Build a Habitable Planet*, 458–63; and Kump, Kasting, and Crane, *Earth System*, 159, 225–29.

82. See Anthony D. Barnosky et al., "Has the Earth's Sixth Mass Extinction Already Arrived?" *Nature* 471, 7336 (March 3, 2011), 51–57. It should be noted that the calculations on species extinction in this paper were arrived at without factoring in climate change.

83. Langmuir and Broecker, *How to Build a Habitable Planet*, 589–95.

84. See William E. Connolly, *Facing the Planetary: Entangled Humanism and the Politics of Swarming* (Durham, NC: Duke University Press, 2017).

85. Langmuir and Broecker, *How to Build a Habitable Planet*, 593. In pointing to the importance of biodiversity to agriculture, Kump, Kasting, and Crane point out that the real question here is biodiversity and not simply whether the world can feed seven, nine, or twelve billion people: "the potential problem with modern agriculture is not that it is not productive enough but that it is *uniform*." Kump, Kasting, and Crane, *Earth System*, 374.

86. Connolly, *Facing the Planetary*, 4.

87. Langmuir and Broecker, *How to Build a Habitable Planet*, 580.

88. The distinction I want to draw here *does not quite derive from* but is inspired by Heidegger's discussion of Plato that makes particular reference to the famous allegory of the cave discussed in the *Republic*; see Heidegger, "Plato's Doctrine of Truth," trans. Thomas Sheehan, in *Pathmarks*, trans. Sheehan et al., ed. William McNeill (1967; New York: Cambridge University Press, 1998) 136–54.

89. See Geoff Mann and Joel Wainwright, "Planetary Sovereignty," *Climate Leviathan: A Political Theory of Our Planetary Future* (New York: Verso, 2018), 129–56. "Planetary sovereignty" refers here to some kind of world government or world order that will manage global warming. See also Duncan Kelly, *Politics and the Anthropocene* (Cambridge: Polity, 2019).

90. Quentin Meillassoux, *After Finitude: An Essay on the Necessity of Contingency*, trans. Ray Brassier (New York: Continuum, 2009), 10.

91. Carl Schmitt, "Dialogue on New Space," *Dialogues on Power and Space*, ed. Andreas Kalyvas and Frederico Finchelstein, trans. and with an introduction by Samuel Garrett Zeitlin (1958; Cambridge: Polity, 2015), 73, 74.

92. Ellis, *Anthropocene*, 144.

93. Ellis, 157.

94. Lenton, *Earth System Science*, 107, 117.

95. Langmuir and Broecker, *How to Build a Habitable Planet*, 645.

96. Langmuir and Broecker, 650, emphasis in original.

97. Langmuir and Broecker, 599–600.

98. Langmuir and Broecker, 668.

99. Timothy M. Lenton and Bruno Latour, "Gaia 2.0: Could Humans Add Some Level of Self-Awareness to Earth's Self-Regulation," *Science*, September 14, 2018, 1068, emphasis added, https://science.sciencemag.org/content/sci/361/6407/1066.full.pdf.

100. "The political world is a pluriverse, not a universe." Carl Schmitt, *The Con-*

cept of the Political, trans. George Schwab (1932; Chicago: University of Chicago Press, 2007), 53.

101. On all this, see Mark Williams et al., "The Anthropocene Biosphere," *Anthropocene Review* 2, no. 3 (2015): 196–219. One has to remember that even the weak 2015 Paris agreement between nations *simply assumes* that toward the end of this century humans will have the technology to draw down CO_2 from the atmosphere—that is, produce "negative" emissions. See Johan Rockström et al., "The World's Biggest Gamble," *Earth's Future* 4 (2016): 465–70, and Oliver Geden, "The Paris Agreement and the Inherent Inconsistency of Climate Policy Making," *WIREs Climate Change* 7 (November/December 2016): 790–97.

102. Félix Guattari, *The Three Ecologies*, trans. Ian Pindar and Paul Sutton (New Brunswick, NJ: Athlone, 2000), 66. Guattari, however, was prophetic about the rise of "men like Donald Trump" in the world he analyzed (43).

103. Latour's name, of course, has to be invoked here as one of the pioneers of this argument. For discussion of Guattari's views, see Jane Bennett, *Vibrant Matter: A Political Ecology of Things* (Durham, NC: Duke University Press, 2010), 113.

104. I owe this point to discussions with Norman Wirzba, whom I thank for sharing his unpublished essay, "Rethinking the Human in an Anthropocene World." What I say here also resonates with some remarks that Joyce Chaplin has recently made: "The term Anthropocene . . . simultaneously promote[s] and diminish[es] humankind. . . . Our collective acts constitute a Great Acceleration. . . . Hurrah for us? Not really. The net result has been a vast reminder that we are just another species, . . . dependent on natural resources for our flourishing and are vulnerable when those . . . become scarce." Joyce Chaplin, "Can the Nonhuman Speak? Breaking the Chain of Being in the Anthropocene," *Journal of the History of Ideas* 78 (October 2017): 512.

105. See chapter 5 below.

106. Holly Robertson, "Snipers to cull up to 10,000 camels in drought-stricken Australia," *Phys.Org*, January 8, 2020, https://phys.org/news/2020-01-snipers-cull-camels-drought-stricken-australia.html.

107. Latour spells out some of his thoughts on this question in Bruno Latour, *Down to Earth: Politics in the New Climatic Regime*, trans. Porter (Medford, MA: Polity, 2018).

108. Bennett, *Vibrant Matter*, 119.

109. Bennett, 13. Kelly Oliver's attempt to develop an earth ethic out of the philosophy of Heidegger is somewhat similar in spirit; see Kelly Oliver, "The Earth's Refusal: Heidegger," in *Earth and World: Philosophy after the Apollo Missions* (New York: Columbia University Press, 2015).

110. Zalasiewicz, *The Earth after Us: What Legacy Will Humans Leave in the Rocks?* (New York: Oxford University Press, 2008), 1.

111. "And then I turned to myself and asked, 'Who are you?'" Saint Augustine, *Confessions*, trans. R. S. Pine-Coffin (London: Penguin, 1961), 212.

CHAPTER FOUR

1. Michael B. Gerrard and Gregory E. Wannier, eds., *Threatened Island Nations: Legal Implications of Rising Seas and a Changing Climate* (Cambridge: Cambridge University Press, 2013), xvii, cited in Edvard Hviding, "Climate Change, Oceanic Sovereignties and Maritime Economies in the Pacific" (paper presented in Oceanic Anthropology

Lecture Series, University of Hawai'i at Manoa, Center for Pacific Islands Studies, and East-West Center, Pacific Islands Development Program, February 13, 2017). I am grateful to Professor Hviding for sharing this paper with me.

2. Anil Agarwal and Sunita Narain, *Global Warming in an Unequal World: A Case of Environmental Colonialism* (1991; repr., New Delhi: Centre for Science and Environment, 2003), 20n1.

3. Agarwal and Narain, 1.

4. François Hartog, *Regimes of Historicity: Presentism and Experiences of Time*, trans. Saskia Brown (New York: Columbia University Press, 2015). Hartog tells a European story of a modern "regime of historicity" (a vista of an open future) that spanned the eighteenth and nineteenth centuries in Europe and came to an end with the two world wars and succumbed to "presentism"—future collapsing into the present—at the end of the twentieth century. One could argue that a similar regime of modern historicity was initiated outside of Europe from the 1950s when decolonizing new nations fell under the spell of modernization theories emanating from both the Soviet Union and the United Stated during the era of the Cold War.

5. Agarwal and Narain, *Global Warming in an Unequal World*, 1.

6. Ellen Barry and Carol Davenport, "Emerging Climate Accord Could Push A/C out of Sweltering India's Reach," *New York Times*, October 12, 2016, https://www.nytimes.com/2016/10/13/world/asia/india-air-conditioning.html.

7. Michael Greenstone, "India's Air-Conditioning and Climate Change Quandary," *New York Times*, October 26, 2016, https://www.nytimes.com/2016/10/27/upshot/indias-air-conditioning-and-climate-change-quandary.html?_r=0.

8. Barry and Davenport, "Emerging Climate Accord."

9. See Ranajit Guha, "The Small Voice of History," in *The Small Voice of History: Collected Essays* (New Delhi: Permanent Black, 2009).

10. Anthony D. Barnosky and Elizabeth A. Hadly, *End Game: Tipping Points for Planet Earth* (London: William Collins, 2015), 41.

11. Barnosky and Hadly, 50–51.

12. Barnosky and Hadly, 43. See also the statistics provided in Robin Jeffrey and Assa Doron, *Waste of a Nation* (Cambridge, MA: Harvard University Press, 2018), 47–54.

13. Barry and Davenport, "Emerging Climate Accord."

14. Arjun Appadurai, *The Future as a Cultural Fact: Essays on the Global Condition* (London: Verso, 2013), 187.

15. Barry and Davenport, "Emerging Climate Accord."

16. Greenstone, "India's Air-conditioning."

17. Elizabeth A. Povinelli, *Geontologies: A Requiem to Late Liberalism* (Durham, NC: Duke University Press, 2016); Déborah Danowski and Viveiros de Castro, *The Ends of the World* (Cambridge: Polity, 2016); William E. Connolly, *Facing the Planetary: Entangled Humanism and the Politics of Swarming* (Durham, NC: Duke University Press, 2017); Michael S. Northcott, *A Political Theology of Climate Change* (Grand Rapids, MI: William B. Eerdmans, 2013).

18. Jane Bennett, *Vibrant Matter: A Political Ecology of Things* (Durham, NC: Duke University Press, 2010), 108.

19. See the poem "Kabi" (The poet) in *Premendra Mitra Rachanabali* [The collected

works of Premendra Mitra] (Calcutta: Granthalaya, 1976), 7. The poem was probably first published in the 1920s and then reprinted in Mitra's first collection of poems, *Prathama* (1932). The contemporary popularity of these lines is attested to by the admiring reference to them in the celebrated conversations of the reputed Bengali polyglot and intellectual Benoykumar Sarkar in his conversations of April 5, 1944. See *Benoy Sarkarer Baithoke* (*Bingsho shotabdir bongo sanskriti*) [in Bengali], vol. 2, conversations with Haridas Mukhopadhyay, Shibchandra Datta, Hemendrabijoy Sen, Kshiti Mukhopadhyay, Subodhkrishna Ghoshal, and Manmathanath Sarkar (Calcutta: Dey's, 2003), 584. The poem was also part of my high school syllabus in the 1960s.

20. See, for example, Andreas Malm, *Fossil Capital: The Rise of Steam Power and the Roots of Global Warming* (New York: Verso, 2016).

21. Bruno Latour, "*Onus orbis terrarum*: About a Possible Shift in the Definition of Sovereignty," *Millennium: Journal of International Studies* 44, no. 3 (2016): 318. Philippe Descola grants similar "advantages" to the nature/culture opposition: "I am ready to concede that such a prison (nature-culture opposition) does have its advantages. Dualism is not an evil in itself and it is ingenuous to stigmatize it for purely moral reasons in the manner of ecologically friendly philosophies of the environment or to blame it for all the evils of the modern era, ranging from colonial expansion to the destruction of nonrenewable resources including the reification of sexual identities and class distinctions. We need at least to give dualism credit not only for its wager that nature is subject to laws of its own but also for its formidable stimulation of the development of the natural sciences. We are also indebted to it not only for the belief that humanity becomes gradually civilized by increasing its own control over nature and disciplining its instincts more efficiently but also for certain *advantages*, in particular political ones, engendered by an aspiration toward progress." Philippe Descola, *Beyond Nature and Culture*, trans. Janet Lloyd (Chicago: University of Chicago Press, 2013), 80–1, emphasis added.

22. Jan Zalasiewicz, Colin N. Waters, Mark Williams, and Colin P. Summerhayes eds., *The Anthropocene as a Geological Time Unit: A Guide to the Scientific Evidence and Current Debate* (Cambridge: Cambridge University Press, 2019), 44.

23. Zalasiewicz et al., *Anthropocene as a Geological Time Unit*, 44. For an extended consideration of the hybrid nature of the materiality that marks "modern" life, see Kylie Ann Crane, "Concrete and Plastic: Thinking through Materiality" (Habilitationsschrift, Universität Potsdam, 2019). I am grateful to Dr. Crane for sharing with me her Habilitation thesis.

24. Zalasiewicz et al., 43.

25. Amartya Sen, *Development as Freedom* (New York: Anchor Books, 2000, and *On Ethics and Economics* (New York: Basil Blackwell, 1987).

26. Bruno Latour, *We Have Never Been Modern*, trans. Catherine Porter (Cambridge, MA: Harvard University Press, 1993. Latour's argument receives a succinct summary in Philippe Descola's brilliant and thoughtful book, *Beyond Nature and Culture*, 86.

27. Bruno Latour, *An Inquiry Into Modes of Existence* (Cambridge, MA: Harvard University Press, 2013).

28. Bruno Latour, *The Pasteurization of France*, trans. Catherine Porter (Cambridge, MA: Harvard University Press, 1988); *Politics of Nature: How to Bring the Sciences into*

Democracy, trans. Catherine Porter (1999; Cambridge, MA: Harvard University Press, 2004). See also the discussion in chapter 6 below.

29. Bruno Latour, *Facing Gaia: Eight Lectures on the New Climatic Regime*, trans. Catherine Porter (Cambridge: Polity, 2017).

30. Descola, *Beyond Nature and Culture*, 87.

31. Latour, *We Have Never Been Modern*, 9, 31; and his *War of the Worlds: What about Peace?* (Chicago: Prickly Paradigm Press, 2002), 31. See also Descola, *Beyond Nature and Culture*, chap. 1.

32. Bruno Latour, Gifford Lectures, Lecture 6, "Gaia's Estate," privately circulated, 126.

33. Aimé Césaire, *Discourse on Colonialism*, trans. Joan Pinkham (New York: Monthly Review Press, 1972), 25.

34. I owe this polemical phrase to Dr. Maira Hayat, who researches the politics of water consumption in contemporary Pakistan.

35. On the history of global concerns around food security for growing number of humans in the 1930s and 1940s, see Alison Bashford's *Global Population: History, Geopolitics, and Life on Earth* (New York: Columbia University Press, 2014).

36. Jawaharlal Nehru, public address "in Hindustani," Calcutta, July 14, 1949, *Selected Works of Jawaharlal Nehru*, 2nd ser., vol. 12, edited by S. Gopal et al. (New Delhi: Jawaharlal Nehru Memorial Fund, 1991), 241.

37. Nehru, speech to the Industries Conference, December 18, 1947, in *Independence and After: A Collection of the More Important Speeches of Jawaharlal Nehru from September 1946 to May 1949* (New Delhi: Publications Division, Government of India, 1949), 155.

38. Jawaharlal Nehru, *Glimpses of World History* (Gurgaon: Penguin, 2004); Nehru, *The Discovery of India* (New York: Oxford University Press, 1989).

39. For more on this, see chapter 7 below.

40. Nehru, speech to the Nineteenth Annual Meeting of the Central Board of Irrigation, December 5, 1948, in *Independence and After*, 386. See also Nehru's convocation address at Roorkee University, November 25, 1949 (All India Radio tapes, Nehru Memorial Museum and Library Extracts), which repeats the point about the Himalayas and how various schemes associated with the mountains and their rivers would produce "power, as well as water and canals and irrigation and more food": Jawaharlal Nehru, *Selected Works of Jawaharlal Nehru*, 2nd ser., vol. 14, pt. 1, edited by S. Gopal et al. (New Delhi: Jawaharlal Nehru Memorial Fund, 1992), 227.

41. Baldev Singh, ed., *Jawaharlal Nehru on Science and Society: A Collection of His Writings and Speeches* (New Delhi: Nehru Memorial Museum and Library, 1988), 94–95.

42. Nehru, *Independence and After*, 152.

43. Address delivered at the opening ceremony of the Central Salt Research Institute, Bhavnagar (Saurashtra) on April 10, 1954 in Singh, *Nehru on Science and Society*, 120.

44. Nehru, *Independence and After*, 385–86.

45. Nehru, inaugural address at the Twenty-Ninth Annual Meeting of the Central Board of Irrigation and Power, New Delhi, November 17, 1958, in Singh, *Nehru on Science and Society*, 173.

46. Nehru, speech to the Nineteenth Annual Meeting of the Central Board of Irrigation, New Delhi, December 5, 1948, in Nehru, *Independence and After*, 391.

47. Nehru, inaugural address at the Twenty-Third Annual Meeting celebrating the silver jubilee of the Central Board of Irrigation and Power, New Delhi, November 17, 1952, in Singh, *Jawaharlal Nehru on Science and Society*, 94–95.

48. Singh, 99.

49. Narendra Modi, *Convenient Action: Gujarat's Response to Challenges of Climate Change* (New Delhi: Macmillan, 2011).

50. See Steve Howard, foreword to Modi, *Convenient Action*. Howard is chief executive officer, The Climate Group, London.

51. See the two chapters titled "Powergudas of Gujarat (Small Is Beautiful)" and "Big Is Also Beautiful (Sardar Sarovar Project)" in Modi, *Convenient Action*, 43–64, 66–82.

52. Modi, *Convenient Action*, 13.

53. See Partha Chatterjee, *Nationalist Thought and the Colonial World: A Derivative Discourse?* (London: Zed, 1986); Homi K. Bhabha, *The Location of Culture* (London: Routledge, 1994).

54. Latour, "*Onus orbis terrarum*."

55. Address at the Twentieth Annual Meeting of the Engineering Association of India, New Delhi, December 28, 1962, in Singh, *Nehru on Science and Society*, 241.

56. See Arjun Appadurai, "The Capacity to Aspire: Culture and the Terms of Recognition," in *The Future as a Cultural Fact: Essays on the Global Condition* (London: Verso, 2013), 179–95.

57. A. C. Pigou, *The Economics of Welfare*, 3rd ed. (London: Macmillan, 1929), vii. There is an element of legitimate continuity between this statement and the following one on the question of caring for "the masses" that comes from the pen of a preeminent economist of our times: "The capabilities of the Chinese masses are now immensely superior in many vital respects than those of the Indian masses. They live a good deal longer, have much safer infancy and childhood, can deal more effectively with illness and disease, can mostly read and write and so on." Amartya Sen, *Commodities and Capabilities* (New Delhi: Oxford University Press, 2003; 1987), app. A, p. 50.

58. See Theodore W. Schultz, "Investment in Man: An Economist's View," *Social Service Review* 33, no. 2 (1959): 109–17; Amartya Sen, "Freedom and the Foundations of Justice," in Amartya Sen, *Development as Freedom*, 74. See also Elizabeth Chatterjee, "The Asian Anthropocene: Electricity and Fossil Developmentalism," *Journal of Asian Studies* 79, no. 1 (February 2020): 1–22.

CHAPTER FIVE

1. Vemula's suicide note as reproduced in the *Times of India*, January 19, 2016.

2. Vemula's Facebook post of November 10, 2016. "Remembering Rohith Vermula" Facebook page available at https://www.facebook.com/rohith352?ref=br_rs.

3. Sverre Raffnsøe, *Philosophy of the Anthropocene: The Human Turn* (Basingstoke: Palgrave Macmillan, 2016), 10

4. Immanuel Kant, *Critique of Practical Reason*, ed. and trans. Lewis White Beck, 3rd ed. (1956; New York: Macmillan, 1993), 169–70.

5. Martha C. Nussbaum, *Hiding from Humanity: Disgust, Shame, and the Law* (Princeton, NJ: Princeton University Press, 2004).

6. Nussbaum, 14. See also 72, 83, 89, 91–93, 94–95, 116.

7. Nussbaum, 17.

8. Nussbaum, 23–24, 50.

9. Nussbaum, 61.

10. Paul G. Falkowski, *Life's Engine: How Microbes Made Earth Inhabitable* (Princeton, NJ: Princeton University Press, 2015); Martin J. Blaser, *The Missing Microbes: How Killing Bacteria Creates Modern Plagues* (New York: Henry Holt, 2014); Eugene G. Grosch and Robert M. Hazen, "Microbes, Mineral Evolution, and the Rise of Microcontinents: Origin and Coevolution of Life with Early Earth," *Astrobiology* 15, no. 10 (2015), https://www.liebertpub.com/doi/10.1089/ast.2015.1302#.

11. See, for instance, Déborah Danowski and Eduardo Viveiros de Castro, *The Ends of the World* (Cambridge: Polity, 2016), on these questions.

12. Sundar Sarukkai, "Phenomenology of Untouchability," in *The Cracked Mirror: An Indian Debate on Experience and Theory*, ed. Gopal Guru and Sundar Sarukkai (New Delhi: Oxford University Press, 2014), 157–99.

13. See Ranajit Guha, *Elementary Aspects of Peasant Insurgency in Colonial India* (New Delhi: Oxford University Press, 1983), chap. 2. See also Kancha Ilaiah, "Productive Labour, Consciousness and History: The Dalitbahujan Alternative," in *Subaltern Studies: Writings on South Asian History and Society*, ed. Shahid Amin and Dipesh Chakrabarty (Delhi: Oxford University Press, 1996), 165–200. Ilaiah began by saying, "Mainstream historiography has done nothing to incorporate the Dalitbahujan perspective in the writing of Indian history: *Subaltern Studies* is no exception to this."

14. The late professor M. N. Srinivas acted as pioneer in providing some of the basic conceptual tools—such as "Sanskritization"—deployed in this literature. Andre Beteille is, of course, another intellectual stalwart of this period.

15. See Ciba Foundation, *Caste and Race: Comparative Approaches*, ed. Anthony de Reuck and Julie Knight (Boston: Little, Brown, 1967).

16. John Stratton Hawley, *A Storm of Songs: India and the Idea of the Bhakti Movement* (Cambridge, MA: Harvard University Press, 2015), 24.

17. Hans-Georg Gadamer, *Truth and Method* (London: Sheed and Ward, 1979), 239–40.

18. The first page of the inaugural issue of *Harijan* magazine launched by Gandhi to tackle the problem of "untouchability" carried a translation of this poem by Rabindranath Tagore. See *Harijan* (Poona), 1, no. 1 (February 11, 1933): 1, https://www.gandhiheritageportal.org/journals-by-gandhiji/harijan.

19. Oliver Mendelsohn and Marika Vicziany, *The Untouchables: Subordination, Poverty and the State in Modern India* (Cambridge: Cambridge University Press, 1998), 6.

20. A. Shukra, "Caste: A Personal Perspective," in *Contextualising Caste: Post-Dumontian Approaches*, ed. Mary Searle-Chatterjee and Ursula Sharma (Oxford: Blackwell, 1994), 171.

21. Frantz Fanon, *Black Skin, White Masks*, trans. Charles Lam Markmann (1952; New York: Grove Press, 1967), 110. See also the discussion in David Macey, "Adieu foulard, Adieu madras," in *Frantz Fanon's "Black Skin, White Masks": New Interdisciplinary Essays*, ed. Max Silverman (Manchester: Manchester University Press, 2005), 22.

22. Louis Dumont, *Homo Hierarchicus: The Caste System and Its Implications*, trans. Mark Sainsbury, Louis Dumont, and Bhasia Gulaiti, complete rev. English ed. (Chicago: University of Chicago Press, 1980; originally published in French, 1966), 54

23. Gopal Guru, "Experience, Space, and Justice," in Guru and Sarukkai, *Cracked Mirror*, 84–87.

24. Gyan Prakash, *Bonded Histories: Genealogies of Labor Servitude in Colonial India* (Cambridge: Cambridge University Press, 2003).

25. For a summary of the literature on this subject, see *Applied Microbiology: Open Access* 3, no. 2 (2017), https://www.longdom.org/open-access/role-of-microbes-in-human-health-2471-9315-1000131.pdf. See also Julia Adeney Thomas, "History and Biology in the Anthropocene: Problems of Scale, Problems of Value," *American Historical Review*, 119, no. 5 (December 2014): 1587–1607.

26. Neil Shubin, *The Universe Within: The Deep History of the Human Body* (New York: Vintage Books, 2013), 33.

27. Steve Vanderheiden, *Atmospheric Justice: A Political Theory of Climate Change* (Oxford: Oxford University Press, 2008), 6, emphasis added. See also the discussion on p. 79.

28. Vanderheiden, 7, emphasis added.

29. Vanderheiden, 264n8.

30. Vanderheiden, 79, 104.

31. Vanderheiden, 251–52.

32. Philip Pettit, *Republicanism: A Theory of Freedom and Government* (Oxford: Oxford University Press, 1999), 137, emphasis added.

33. See, for instance, Eduardo Kohn, *How Forests Think: Toward an Anthropology beyond the Human* (Berkeley: University of California Press, 2013). Kohn's work is inspired by the work of Eduardo Vivieros de Castro, Philippe Descola, and others.

34. For a valiant effort to bring the two together, see Samantha Frost, *Biocultural Creatures: Toward a New Theory of the Human* (Durham, NC: Duke University Press, 2016).

35. The letter, dated February 29, 1912, is reprinted in Rabindranath Tagore, *Chithipotro* [Letters], ed. Bhabatosh Datta, vol. 15 (Calcutta: Visva-Bharati, 1995), 71–73. The editorial note by Bhabatosh Datta on p. 220 wrongly suggests that the letter was written to Gourhari Sen, a founder of one of the oldest public libraries in Calcutta, the Chaitanya Library (1889). The background to this correspondence may be found in Prasantakumar Pal, *Robijiboni* [Life of Robi], vol. 6 (Calcutta: Ananda, 1993), 247–48. I am grateful to Sanjib Mukhopadhyay for bringing this letter to my attention and for discussion on its historical context.

36. The original letter may be found in Tagore's *Chinnapatrabali* in Rabindranath Thakur [Tagore], *Rabindrarachanabali* [Collected works of Rabindranath], vol. 11 (Calcutta: Government of West Bengal, 1961), 74.

37. Tagore, *Chithipotro*, 15:72.

CHAPTER SIX

1. Tim Lenton, "2°C or not 2°C? That Is the Question," *Nature* 473 7 (May 5, 2011): 7, http://www.nature.com/news/2011/110504/pdf/473007a.pdf. For the exact wording of the phrase, see article 2 of the *United Nations Framework Convention on Climate Change* (New York: United Nations, 1992), 4, https://unfccc.int/resource/docs/convkp/conveng.pdf.

2. Eric Holtaus, "When Will the World Really Be 2 Degrees Hotter Than It Used to

Be?," *FiveThirtyEight*, March 23, 2016, http://fivethirtyeight.com/features/when-will-the-world-really-be-2-degrees-hotter-than-it-used-to-be/. I am grateful to James Chandler for drawing my attention to this article.

3. Jan Zalasiewicz and Mark Williams, *The Goldilocks Planet: The Four Billion Year Story of Earth's Climate* (Oxford: Oxford University Press, 2012).

4. Julia Adeney Thomas, "History and Biology in the Anthropocene: Problems of Scale, Problems of Value," *American Historical Review* 119, no. 5 (December 2014): 1588.

5. Tim Flannery, *The Weather Makers* (Melbourne: Text, 2007), chap. 22, "Civilisation: Out with a Whimper?"

6. Dipesh Chakrabarty, *The Crises of Civilization: Explorations in Global and Planetary Histories* (New Delhi: Oxford University Press, 2018).

7. Peter Singer, "Climate Change: Our Greatest Ethical Challenge," lecture at the University of Chicago, October 23, 2015.

8. For more on this, see my "The Future of the Human Sciences in the Age of Humans: A Note," *European Journal of Social Theory* 20, no. 1 (2017): 39–43.

9. Martha C. Nussbaum, *Not for Profit: Why Democracy Needs the Humanities* (Princeton, NJ: Princeton University Press, 2012), 7.

10. Sue Donaldson and Will Kymlicka, *Zoopolis: A Political Theory of Animal Rights* (New York: Oxford University Press, 2011), 15 and chap. 2 generally; Robert Garner, *The Political Theory of Animal Rights* (Manchester: Manchester University Press, 2005), 14–15, 125–28; Robert Garner, *A Theory of Justice for Animals: Animal Rights in a Nonideal World* (New York: Oxford University Press, 2013), 3, 133. For a stimulating discussion of the philosophical-moral dilemmas and problems encountered in this area of study, see Cary Wolfe, *Before the Law: Humans and Other Animals in a Biopolitical Frame* (Chicago: University of Chicago Press, 2013).

11. In the following couple of paragraphs, I draw on my "response" piece in Robert Emmett and Thomas Lekan, eds., "Whose Anthropocene? Revisiting Dipesh Chakrabarty's 'Four Theses,'" *RCC Perspectives: Transformations in Environment and Society* 2016, no. 2 (2016), http://www.environmentandsociety.org/perspectives/2016/2/whose-anthropocene-revisiting-dipesh-chakrabartys-four-theses.

12. Hans Jonas, *The Imperative of Responsibility: In Search of an Ethics for the Technological Age*, trans. Hans Jonas with David Herr (Chicago: University of Chicago Press, 1984), 22.

13. See, for instance, Steve Vanderheiden, *Atmospheric Justice: A Political Theory of Climate Change* (New York: Oxford University Press, 2008).

14. See chapters 1 and 2 above.

15. John L. Brooke, *Climate Change and the Course of Global History* (New York: Cambridge University Press, 2014), 121–22.

16. Dipesh Chakrabarty, "The Human Significance of the Anthropocene," in *Modernity Reset!*, ed. Bruno Latour (Cambridge, MA: MIT Press, 2016), 189–99. I have benefited from discussions with Henning Trüper on this point.

17. Daniel Lord Smail, *Deep History and the Brain* (Berkeley: University of California Press, 2007).

18. Yuval Noah Harari, *Sapiens: A Brief History of Humankind* (New York: Harper Collins, 2015), 11.

19. Harari, 11.

20. Harari, 11-12. A similar point is made by Hans Jonas while comparing the speed of human-technological changes to that of changes brought about by natural evolution: "Natural evolution works with small things, never plays for the whole stake, and therefore can afford innumerable 'mistakes' in its single moves, from which its patient, slow process chooses the few, equally small 'hits.' . . . Modern technology, neither patient nor slow, compresses . . . the many infinitesimal steps of natural evolution into a few colossal ones and foregoes by that procedure the vital advantages of nature's 'playing safe.'" Jonas, *Imperative of Responsibility*, 31. See also the section on "Man's Disturbance of the Symbiotic Balance," 138.

21. Harari, *Sapiens*, 11-12.

22. The last big famine India saw, for example, was in 1943, though many in the country still die of hunger and malnutrition.

23. William R. Catton Jr., *Overshoot: The Ecological Basis of Revolutionary Change* (Chicago: University of Illinois Press, 1980), 95-96; D. Cocks, *Global Overshoot: Contemplating the World's Converging Problems* (New York: Springer, 2013).

24. See Lewis Regenstein, "Animal Rights, Endangered Species and Human Survival," in *In Defense of Animals*, ed. Peter Singer (New York: Basil Blackwell, 1987), 118-32; Peter Singer, "Down on the Factory Farm," in *Animal Rights and Human Obligations*, ed. Tom Regan and Peter Singer (1976; Englewood Cliffs, NJ: Prentice Hall, 1989), 159-68.

25. See Jessica C. Stanton et al., "Warning Times for Species Extinction due to Climate Change," *Global Change Biology* 21 (2015): 1066-77; Rodolfo Dirzo et al., "Defaunation in the Anthropocene," *Science* 345, no. 6195 (2014): 401-6; Celine Bellard et al., "Impacts of Climate Change on the Future of Biodiversity," *Ecology Letters* 15, no. 4 (April 2012): 365-77; Gerardo Ceballos et al., "Accelerated Modern Human-Induced Species Losses: Entering the Sixth Mass Extinction," *Science Advances*, June 19, 2015, 1-5.

26. Will Steffen, Wendy Broadgate, Lisa Deutsch, Owen Gaffney, and Cornelia Ludwig, "The Trajectory of the Anthropocene: The Great Acceleration," *Anthropocene Review* 2, no. 1 (2015): 1-18.

27. Steffen et al., 1-18.

28. Rockström, J., "Planetary boundaries: exploring the safe operating space for humanity," *Ecology and Society* 14, no. 2 (2009): 32, http://www.ecologyandsociety.org/vol14/iss2/art32/.

29. Richard E. Zeebe et al., "Anthropogenic Carbon Release Rate Unprecedented during the Past 66 Million Years," *Nature Geoscience* 9 (March 21, 2016): 325-29, https://www.nature.com/articles/ngeo2681.

30. "If global warming and a sixth extinction take place in the next couple of centuries, then an epoch will seem too low a category in the hierarchy [of the geological timetable]." Jan Zalasiewicz, personal communication with the author, September 30, 2015.

31. Gary Tomlinson, "Two Deep-Historical Models of Climate Crisis," *South Atlantic Quarterly* 116, no. 1 (January 2017): 19-31.

32. Jan Zalasiewicz, "The Geology behind the Anthropocene" (unpublished typescript, 2015), 12. I am grateful to Zalasiewicz for sharing this paper with me.

33. See the detailed and excellent discussion in Frank P. Incropera, *Climate Change:*

A Wicked Problem—Complexity and Uncertainty at the Intersection of Science, Economics, Politics, and Human Behavior (New York: Cambridge University Press, 2016).

34. The separation was formalized in the nineteenth century when the modern social sciences arose as an identifiable set of disciplines. See Fabien Locher and Jean-Baptiste Fressoz, "Modernity's Frail Climate," *Critical Inquiry* 38, no. 3 (Spring 2012): 579–98.

35. The saga of that intellectual war is recapitulated in Eric M. Gander, *On Our Minds: How Evolutionary Psychology Is Reshaping the Nature-versus-Nurture Debate* (Baltimore: Johns Hopkins University Press, 2003), chap. 3.

36. For a different, stimulating, and critical reading of this essay, see Bonnie Honig, *Political Theory and the Displacement of Politics* (Ithaca, NY: Cornell University Press, 1993), 19–24.

37. Immanuel Kant, "Speculative Beginning of Human History (1786)," in *Perpetual Peace and Other Essays on Politics, History, and Morals*, trans. and with an introduction by Ted Humphrey (Indianapolis, IN: Hackett, 1988), 54. The words within brackets belong to this edition of the essay.

38. Kant, "Speculative Beginning of Human History," 54–55. For the use of *animal*, *natural*, and *physical* as synonyms, see 54n.

39. See the literature discussed in and the conclusions of Daniel P. Sheilds, "Aquinas and the Kantian Principle of Treating Persons as Ends in Themselves," (PhD diss., Catholic University of America, 2012), chaps, 1, 2.

40. I completely agree with Honig's remark that "the stories fables tell about the founding of a form of life invariably serve as powerful illustrations of the now more subtle and sedimented but no less active processes and practices that constitute and maintain our present, daily." Honig, *Political Theory*, 19.

41. Kant, "Speculative Beginning of Human History," 49, emphasis added.

42. Kant, 49–50, emphasis in original.

43. Kant, 50–51, emphases in original.

44. Kant, 51.

45. Kant, 52, emphasis in original.

46. Kant, 52–53, emphasis in original.

47. See, again, Honig, *Political Theory*, 27–34, for a rich and complex reading of these Kantian injunctions.

48. Kant, "Speculative Beginning of Human History," 53.

49. Kant, 53–54.

50. Kant, 56–57.

51. Kant, 57.

52. Kant, 54.

53. Kant, 57–58.

54. Kant, 57–59.

55. Kant, 59.

56. Immanuel Kant, *The Critique of Judgement*, trans. James Creed Meredith (1928; Oxford: Clarendon, 1973; first published in German, 1790), pt. 2, 93–94.

57. Kant, "Speculative Beginning of Human History," 55n.

58. Kant, 51.

59. For a critical discussion of some of the issues involved here, see David Bau-

meister's critical but generous reading of the essay on which this chapter is based. David Baumeister, "Kant, Chakrabarty, and the Crisis of the Anthropocene," *Environmental Ethics* 41, no. 1 (2019): 53–67. I agree with his point that my critique does not exhaust the role that reason could play in the current crisis. See also in this context Clive Hamilton, "The Delusion of the 'Good Anthropocene': Reply to Andrew Revkin," June 14, 2014, http://clivehamilton.com/the-delusion-of-the-good-anthropocene-reply-to-andrew-revkin/.

60. Pope Francis, *Encyclical on Climate Change and Inequality: On Care for Our Common Home*, introduction by Naomi Oreskes (Brooklyn, NY: Melville House, 2015), 72–74.

61. Pope Francis, 73.

62. Pope Francis, 42, 43.

63. Amartya Sen, "Energy, Environment, and Freedom: Why We Must Think about More Than Climate Change," *New Republic*, August 25, 2014, 39.

64. Peter F. Sale, *Our Dying Planet: An Ecologist's View of the Crisis We Face* (Berkeley: University of California Press, 2011), 223. J. R. McNeill and Peter Engelke note in their book *The Great Acceleration: An Environmental History of the Anthropocene since 1945* (Cambridge, MA: Harvard University Press, 2014), 87, that of the quarter of a million species that went extinct in the twentieth century, most "disappeared before they could be described by scientists" and were creatures "unknown to biology."

65. Martin J. Blaser, *Missing Microbes: How the Overuse of Antibiotics Is Fueling Our Modern Plagues* (New York: Picador, 2014), 13–14, 15, 16.

66. Blaser, chap. 9.

67. Luis P. Villarreal, "Can Viruses Make Us Human?," *Proceedings of the American Philosophical Society* 148, no. 3 (September 2004): 296–323; Linda M. van Blerkom, "Role of Viruses in Human Evolution," *Yearbook of Physical Anthropology* 46 (2003): 14–46.

68. See the discussion in chapter 2 above.

69. See in particular Bruno Latour, *We Have Never Been Modern*, trans. Catherine Porter (Cambridge, MA: Harvard University Press, 1993); *Politics of Nature: How to Bring the Sciences into Democracy*, trans. Catherine Porter (1999; Cambridge, MA: Harvard University Press, 2004); *An Inquiry into Modes of Existence: An Anthropology of the Moderns*, trans. Catherine Porter (Cambridge, MA: Harvard University Press, 2013).

70. Bruno Latour, *The Pasteurization of France*, trans. Alan Sheridan and John Law (Cambridge, MA: Harvard University Press, 1993), 193; *Politics of Nature*; *Inquiry into Modes of Existence*.

71. Latour, *Pasteurization of France*, 32, 33.

72. Latour, 35.

73. Latour, 33–34.

74. Latour, 39, 43.

75. Latour, 149–50.

76. Latour, 150.

77. Blaser, *Missing Microbes*, 12–13.

78. Latour, *Pasteurization of France*, 193.

79. Latour, 193.

80. Latour, 192–94. Claire Colebrook asked—in an essay engaging Derrida in the context of a discussion on the Anthropocene—a question that reverberates through

much posthumanist thinking: "Is not the notion that the earth is *our place* precisely what has blinded us to the ravages of our mode of life?" Claire Colebrook, "Not Symbiosis, Not Now: Why Anthropogenic Change Is Not Really Human," *Oxford Literary Review* 34, no. 2 (2012): 189.

CHAPTER SEVEN

1. Jan Zalasiewicz, "The Extraordinary Strata of the Anthropocene," in *Environmental Humanities: Voices from the Anthropocene*, ed. S. Oppermann and S. Iovino (London: Rowman and Littlefield International, 2017), 124. I am grateful to Jan Zalasiewicz for sharing this essay with me. My arguments here do not in any way assume or need to assume that the proposal to formalize the Anthropocene will be ratified. I don't think that Zalasiewicz's argument that I use in this chapter makes that assumption either.

2. Zalasiewicz, "Extraordinary Strata," 9.

3. John Bellamy Foster, foreword to *Facing the Anthropocene: Fossil Capitalism and the Crisis of the Earth System*, by Ian Angus (New York: Monthly Review Press, 2016), 11.

4. Foster, 11.

5. Will Steffen, commentary on Paul J. Crutzen and Eugene F. Stoermer, "The Anthropocene," in *The Future of Nature: Documents of Global Change*, ed. Libby Robin, Sverker Sörlin, and Paul Warde (New Haven, CT: Yale University Press, 2013), 486.

6. Paul J. Crutzen and Eugene F. Stoermer, "The Anthropocene," *IGBP Newsletter* 41 (2000): 17, cited in chapter 1.

7. David Archer, *The Long Thaw: How Humans Are Changing the Next 100,000 Years of Earth's Climate* (Princeton, NJ: Princeton University Press, 2009), 64.

8. Archer, 6.

9. Will Steffen, Jacques Grinevald, Paul Crutzen, John McNeill, "The Anthropocene: Conceptual and Historical Perspectives," *Philosophical Transactions of the Royal Society A* 369, no. 1938 (2011): 843.

10. John R. McNeill, *Something New Under the Sun: An Environmental History of the Twentieth-Century World* (New York: W. W. Norton, 2000), 3.

11. Roger Revelle and Hans E. Suess, "Carbon Dioxide Exchange between Atmosphere and Ocean and the Question of an Increase of Atmospheric CO_2 during the Past Decades," *Tellus* 9, no. 1 (1957): 18–27, reproduced in *The Warming Papers: The Scientific Foundation for the Climate Change Forecast*, ed. David Archer and Raymond Pierrehumbert (Oxford: Wiley-Blackwell, 2011), 277.

12. Sheila Jasanoff, "A New Climate for Society," *Theory, Culture and Society* 27, no. 2/3 (2010): 236.

13. The most outstanding, original, and learned philology of the term *Anthropocene* to my knowledge is Robert Stockhammer's essay "Philology of the Anthropocene," in *Meteorologies of Modernity: Weather and Climate Discourses in the Anthropocene*, ed. Sarah Fekadu, Hanna Straß-Senol, and Tobias Döring, Yearbook of Research in English and American Literature 33 (Tubingen: Narr, 2017), 43–64.

14. Sometimes, of course, "force" and "power" are used loosely to mean the same thing, but for the sake of clarity of exposition, I will treat them as belonging, respectively, to "natural" and "social" history. This is not an arbitrary distinction. The historical-existential nature of the category "power" is what enables Foucault's nomi-

nalist exercise in describing the nature of power in his *History of Sexuality*, vol. 1, *An Introduction*, trans. Robert Hurley (New York: Vintage Books, 1978), pt. 3, chap. 2, 92–97.

15. *Climate Change: The IPCC Scientific Assessment*, ed. J. T. Houghton, G. J. Jenkins, and J. J. Ephraums (Cambridge: Cambridge University Press, 1991), "Policymakers Summary," xiii.

16. Anil Agarwal and Sunita Narain, *Global Warming in an Unequal World: A Case of Environmental Colonialism* (1991; repr., New Delhi: Centre for Science and Environment, 2003), 1, 20n1.

17. Agarwal and Narain, 1.

18. François Hartog, *Regimes of Historicity: Presentism and Experiences of Time*, trans. Saskia Brown (New York: Columbia University Press, 2015). Hartog, of course, tells a European—though not Eurocentric—story of a modern "regime of historicity" (a vision of an open futural time) in Europe that spanned the eighteenth and nineteenth centuries and came to an end with the two world wars and succumbed to a "presentism"—future collapsing into the present—at the end of the twentieth century. Incidentally, Ursula Heise describes the Anthropocene precisely in terms that are reminiscent of Hartog's description of "presentism"—"as a future that has already arrived." Ursula K. Heise, *Imagining Extinction: The Cultural Meanings of Endangered Species* (Chicago: University of Chicago Press, 2016), 203, 219–20. For a different argument about history's relationship to future—at least in the Western imagination—see Zoltán Boldizsár Simon, *History in Times of Unprecedented Change: A Theory for the 21st Century* (London: Bloomsbury Academic, 2019).

19. Agarwal and Narain, *Global Warming in an Unequal World*, 1.

20. Andreas Malm and Alf Hornborg, "The Geology of Mankind? A Critique of the Anthropocene Narrative," *Anthropocene Review* 1, no. 1 (2014): 66. Matthew Lepori, "There Is No Anthropocene: Climate Change, Species-Talk, and Political Economy," *Telos* 172 (Fall 2015): 103–24, makes similar points.

21. Malm and Hornborg, "The Geology of Mankind?," 64, 66.

22. Malm and Hornborg, 62, emphasis added.

23. Malm and Hornborg, 67.

24. Jason W. Moore, *Capitalism and the Web of Life: Ecology and the Accumulation of Capital* (London: Verso, 2015). Donna Haraway writes that "personal email communication from both Jason Moore and Alf Hornborg in late 2014 told me Malm proposed the term Capitalocene in a seminar in Lund, Sweden, 2009, when he was still a graduate student. I first used the term independently in public lectures in 2012." Haraway, "Anthropocene, Capitalocene, Plantationocene, Chthulucene: Making Kin," *Environmental Humanities* 6, no. 1 (2015): 163n6. However, Christian Schwägerl, *The Anthropocene: The Human Era and How It Shapes Our Planet*, trans Lucy Renner Jones (Santa Fe and London: Synergetic Press, 2014), 65n132, gives an alternative origin for the term: "The term 'Kapitalozän' . . . was coined by Prof. Elmar Altvater from the Freie Universität, Berlin, during a discussion at the German Council of Foreign Relations."

25. Moore, *Capitalism*, 173n13.

26. Moore, 169–71, emphasis added.

27. Ian Angus, *Facing the Anthropocene: Fossil Capitalism and the Crisis of the Earth System* (New York: Monthly Review Press, 2016), 231–32, where he describes Capitalocene as

a "category mistake": "capitalism is a 600-year-old social and economic system, while the Anthropocene is a 60-year-old Earth System epoch.... The new epoch will continue long after capitalism is a distant memory."

28. Angus, 109–10. The move that Malm, Hornborg, Moore, Angus, and others made—of analyzing anthropogenic climate change through inequalities among humans and hence through appeals to theorizations of "culture and power"—is not surprising. This is how many world-history analysts had earlier dealt with global environmental problems and their histories: by focusing on how they were mediated by human inequalities, the rise of "developmentalist projects," and state power in parts of the world in the period 1500–1800 and changing human constructions of nature under conditions of modernity. See, for instance, William Cronon, ed., *Uncommon Ground: Rethinking the Human Place in Nature* (New York: W. W. Norton, 1996); Alf Hornborg, J. R. McNeill, and Joan Martinez-Alier, eds., *Rethinking Environmental History: World-System History and Global Environmental Change* (Lanham, MD: Altamira, 2007); Edmund Burke and Kenneth Pomeranz, eds., *The Environment and World History* (Berkeley: University of California Press, 2009). The expression "developmentalist projects" is Kenneth Pomeranz's. See his introduction to the Burke and Pomeranz volume.

29. Jan Zalasiewicz, Will Steffen, Reinhold Leinfelder, Mark Williams, and Colin Waters, "Petrifying Earth Process: The Stratigraphic Imprint of Key Earth System Parameters in the Anthropocene," *Theory, Culture, Society* 34, no. 2/3 (2017): 98.

30. Archer, *Long Thaw*, 10.

31. Zalasiewicz et al., "Petrifying Earth Process," 16, emphasis added.

32. David Grinspoon, *Earth in Human Hands: Shaping Our Planet's Future* (New York: Grand Central, 2016), 242–43.

33. Daniel Schrag, "Geobiology of the Anthropocene," in *Fundamentals of Geobiology*, ed. Andrew H. Knoll, Donald E. Canfield, and Kurt O. Kornhauser (Oxford: Blackwell, 2012), 434.

34. Clive Hamilton, *Defiant Earth: The Fate of Humans in the Anthropocene* (Cambridge: Polity, 2017), 41, emphasis added.

35. William H. McNeill, "The Changing Shape of World History," in *World History: Ideologies, Structures, and Identities*, ed. Philip Pomper, Richard A. Elphick, and Richard T. Vann (Malden, MA: Blackwell, 1998), 39–40. This view was somewhat different from what another pioneering scholar of world history, Marshall Hodgson, McNeill's colleague at the University of Chicago, thought of very large-scale histories of humanity: "If world history is philosophically possible, it will in any case be subject to two important limitations. It will not only be unlikely to deal with all or even most of the events that have troubled mankind from the beginning; further it is unlikely to bear the type of human meaning which a sensitive history of a particular small community can have." Marshall Hodgson, "The Objectivity of Large-Scale Historical Inquiry: Its Peculiar Limits and Requirements," in *Rethinking World History: Essays on Europe, Islam, and World History*, ed. Edmund Burke III (Cambridge: Cambridge University Press, 1993), 258. For a fascinating account of Hodgson's intellectual background and his interactions with McNeill at the University of Chicago, see Michael Geyer, "Marshall G. S. Hodgson: The Invention of World History from the Spirit of Nonviolent Resistance" (forthcoming). I understand from Professor Geyer that the Hodgson papers archived at the University of Chicago contain "quite a bit of science and some very in-

tense and very critical discussions of Teilhard de Chardin (among other things about the mistake of anthropocentrism)." Personal communication with the author, February 11, 2017.

36. William H. McNeill, "Passing Strange: The Convergence of Evolutionary Science with Scientific History," *History and Theory* 40, no. 1 (2001): 5.

37. McNeill, 15.

38. William H. McNeill, "At the End of an Age?," *History and Theory* 42, no. 2 (2003): 251, 252.

39. John L. Brooke, *Climate Change and the Course of Global History: A Rough Journey* (Cambridge: Cambridge University Press, 2014), 578–79, emphasis added.

40. Reinhart Koselleck, "'Space of Experience' and 'Horizon of Expectation': Two Historical Categories," in *Futures Past: On the Semantics of Historical Time*, trans. Keith Tribe (1979; Cambridge, MA: MIT Press, 1985), 270.

41. Saint Augustine, *Confessions*, trans. R. S. Pine-Coffin (London: Penguin, 1961), 11.20.269.

42. Lucian Hölscher, "Time Gardens: Historical Concepts in Modern Historiography," *History and Theory* 53, no. 4 (2014): 591.

43. Reinhart Koselleck, "Time and History," in *The Practice of Conceptual History: Timing History, Spacing Concepts*, trans. Todd Samuel Presner et al. (Stanford, CA: Stanford University Press, 2002), 110. See also Koselleck's essay, "Concepts of Historical Time and Social History," in the same volume, 115–30, and John Zammito's review essay, "Koselleck's Philosophy of Historical Time(s) and the Practice of History," *History and Theory* 45, no. 1 (2004): 124–35.

44. Koselleck, "'Space of Experience' and 'Horizon of Expectation,'" 272.

45. Koselleck, 274, 275, 284.

46. Koselleck, 275.

47. Koselleck, 274. As Christophe Bouton pointed out in his commentary on Koselleck, "[Koselleck's] . . . categories 'capacity to die and capacity to kill' . . . are a basic transcendental structure of history since, according to Koselleck, the threat of violent death is the background of any history, from the hunter-gatherers to the atomic age. Without the capacity to kill one another, 'the histories we all know would not exist.'" Christophe Bouton, "The Critical Theory of History: Rethinking the Philosophy of History in the Light of Koselleck's Work," *History and Theory* 55, no. 2 (2016): 178.

48. Andrew Light, "Climate Diplomacy," in *The Oxford Handbook of Environmental Ethics*, ed. Stephen Gardiner and Allen Thompson (Oxford: Oxford University Press, 2017).

49. See W. J. T. Mitchell, *The Last Dinosaur Book: The Life and Times of a Cultural Icon* (Chicago: University of Chicago Press, 1998); Bernd Scherer, "Die Monster," in *Das Anthropozän: Zum Stand der Dinge*, ed. Jürgen Renn and Bernd Scherer (Berlin: Matthes und Seitz, 2016), 226–41; Wolfgang Behringer, *Tambora und das Jahr ohne Sommer: Wie ein Vulkan die Welt in die Krise stürzte* (Munich: C. H. Beck, 2016).

50. See my "The Human Significance of the Anthropocene," in *Modernity Reset!*, ed. Bruno Latour (Cambridge, MA: MIT Press, 2016).

51. Zalasiewicz, "Extraordinary Strata," 1.

52. Zalasiewicz, 3. The story of William Smith, "a surveyor in England, who was the first to recognize that fossils added information about the rocks in which they were

found," illustrates Zalasiewicz's point. See David N. Reznick, *The "Origin" Then and Now: An Interpretive Guide to the "Origin of Species"* (Princeton, NJ: Princeton University Press, 2010), 268. See also "The Carboniferous Period," University of California Museum of Paleontology, http://www.ucmp.berkeley.edu/carboniferous/carboniferous.php.

53. Kathleen D. Morrison, "Provincializing the Anthropocene," *Seminar* 673 (September 2015): 75.

54. Morrison, 79.

55. Simon L. Lewis and Mark A. Maslin, "Defining the Anthropocene," *Nature* 519, no. 7542 (2015): 171.

56. Lewis and Maslin, 175, 176.

57. Lewis and Maslin, 177. Text quoted within brackets is from p. 176.

58. Lewis and Maslin, 175.

59. See Clive Hamilton, "Getting the Anthropocene So Wrong," *Anthropocene Review* 2, no. 1 (2015): 1–6; Jan Zalasiewicz et al., "Colonization of the Americas, 'Little Ice Age' Climate, and Bomb-Produced Carbon: Their Role in Defining the Anthropocene," *Anthropocene Review* 2, no. 2 (2015): 117–27; Simon L. Lewis and Mark A Maslin, "A Transparent Framework for Defining the Anthropocene Epoch," *Anthropocene Review* 2, no. 2 (2015): 128–46.

60. Zalasiewicz, "Extraordinary Strata," 3.

61. Zalasiewicz, 4.

62. For a recent statement, see Zalasiewicz et al., "Petrifying Earth Process."

63. The names of geological periods usually have little to do with the factors that may have brought them into being. Thus name *Cretaceous* is "from the Latin word for chalk," *Jurassic* "after the Jura hills on the Franco-Swiss border," *Triassic* "because across much of central Europe it had a tripartite character: two sandstone formations . . . separated by a distinctive limestone," *Silurian* "from the name of an ancient British tribe," *Cambrian* "after the Roman name for Wales," *Denovian* named "after the English county of Devonshire," and so on. Why should not the principle apply to the "stratigraphic Anthropocene"? See Martin J. S. Rudwick, *Earth's Deep History: How It Was Discovered and Why It Matters* (Chicago: University of Chicago Press, 2014), 142–43.

64. Zalasiewicz, "Extraordinary Strata," 11.

65. Zalasiewicz, 9, emphasis added.

66. David Archer, *The Global Carbon Cycle* (Princeton, NJ: Princeton University Press, 2010), 21.

67. Charles H. Langmuir and Wally Broecker, *How to Build a Habitable Planet: The Story of Earth from the Big Bang to Humankind* (Princeton, NJ: Princeton University Press, 2012), 591.

68. Jan Zalasiewicz, personal communication with the author, February 27, 2017. See also Jan Zalasiewicz et al., "Chronostratigraphy and Geochronology: A Proposed Realignment," *GSA Today* 23, no. 3 (2013): 4–8, and Jan Zalasiewicz, Mark Williams, and Colin Waters, "Can an Anthropocene Series Be Defined and Recognized?," *Geological Society, London, Special Publications* 395 (March 2014): 39–53. Bronislaw Szerszynski, "The Anthropocene Monument: On Relating Geological and Human Time," *European Journal of Social Theory* 20, no. 1 (2017): 193, has an illuminating discussion of this point.

69. I am grateful to Fredrik Albritton Jonsson for discussions on this point.

70. St. Augustine, *Concerning the City of God against the Pagans* [1467], trans. Henry Bettenson and introduced by John O'Meara (1972; Harmondsworth: Penguin, 1984), 12.14.13.486; Buffon cited in Paolo Rossi, *The Dark Abyss of Time: The History of the Earth and the History of Nations from Hooke to Vico*, trans. Lydia G. Cochrane (Chicago: University of Chicago Press, 1987), 108; Darwin cited in Pascal Richet, *A Natural History of Time*, trans. John Venerella (Chicago: University of Chicago Press, 2007; first published in French, 1999), 212; Martin J. S. Rudwick, *Worlds before Adam: The Reconstruction of Geohistory in the Age of Reform* (Chicago: University of Chicago Press, 2008), 564. On Darwin's response to the vastness of the deep past, see Joe D. Burchfield, "Darwin and the Dilemma of Geological Time," *Isis* 65, no. 3 (1974): 300–321.

71. The relation between the (modern) time of human history and time of the geological past receives a stimulating discussion in Szerszynski, "The Anthropocene Monument," where he shows how eighteenth- and early nineteenth-century geology "drew on the practices of erudite and antiquarian history . . . to [produce] a *history of the Earth*" (115). There remain differences, however, in methods of constructing historical and geological times: "Historians of human culture have modern examples of revolution or mass hysteria to examine for comparison with records of the past. But, . . . given the complexity of geological events, our lack of experience of all geological environments and of geological spans of time, and our interest in the singularity of each event, geologists simply cannot project the present onto the past." Robert Frodeman, "Geological Reasoning: Geology as an Interpretive and Historical Science," *GSA Bulletin* 107, no. 8 (1995): 965.

72. See chapter 2 above.

73. Langmuir and Broecker, *How to Build a Habitable Planet*, 16–17.

74. Langmuir and Broecker, xv.

75. See, for instance, the very first page of Langmuir and Broecker, *How to Build a Habitable Planet*.

76. Langmuir and Broecker, 650.

77. See Adam Frank and Woodruff Sullivan, "Sustainability and the Astrobiological Perspective: Framing Human Futures in a Planetary Context," *Anthropocene* 5 (March 2014): 32–41.

78. Langmuir and Broecker, *How to Build a Habitable Planet*, 668.

79. Langmuir and Broecker, 645–46, 668.

80. Zalasiewicz, "Extraordinary Strata," 5.

81. Peter Haff, "Technology as a Geological Phenomenon: Implications for Human Well-Being," in *A Stratigraphical Basis for the Anthropocene*, ed. C. N. Waters et al., Geological Society, Special Publications, vol. 395 (London: Geological Society, 2014), 302. This argument makes for a very interesting distinction between intelligence and subjectivity/consciousness. If one thinks of intelligence as a problem-solving property of different forms of life, we find it in life-forms that have no "subjectivity," so to speak; consciousness may be seen as a consequence of the development of the brain. Termites building a mound have to solve some of the same problems of structure that builders of a skyscraper need to address. See the discussion in Andrew Y. Glikson and Colin Groves, *Climate, Fire and Human Evolution: The Deep Time Dimensions of the Anthropocene* (Cham, Switzerland: Springer, 2016), 185–87.

82. Langmuir and Broecker, *How to Build a Habitable Planet*, 645–46.

83. Glikson and Groves, *Climate, Fire and Human Evolution*, 193, 194–95.

84. Glikson and Groves, 194. If human activities eventually lead to a sixth great extinction of species, it will be a "first" for the planet. Never before has a species caused a mass extinction event. They were all caused by "some strong combination of asteroid impacts, clusters of volcanic eruptions, ice ages, and/or indications of large changes in ocean chemistry." Reznick, *The "Origin,"* 310. See also Andrew Glikson and Emily Spence, "Planet Eaters: Chain Reactions, Black Holes, and Climate Change," app. D to Andrew Glikson, *The Event Horizon: Imagining the Real* (Canberra: published by the author, 2016), 92–95. I should also mention here that as a geologist, Glikson prefers to keep the term informal and chooses to speak of early, middle, and late Anthropocene, using the term more as an expression of human impact on the planet rather than as pointing to a specific stratigraphic series.

85. Christophe Bonneuil and Jean-Baptiste Fressoz, *The Shock of the Anthropocene: The Earth, History, and Us*, trans. David Fernbach (London: Verso, 2016), 80.

86. Bonneuil and Fressoz, 94.

87. Mark Lynas, *The God Species* (Washington, DC: National Geographic, 2011).

88. Immanuel Kant, "An Old Question Raised Again: Is the Human Race Constantly Progressing?," in *The Conflict of the Faculties*, trans. Mary J. Gregor (1979; Lincoln: University of Nebraska Press, 1992), 169. Lewis Beck translated this particular essay in 1957 (see publisher's note on the frontispiece).

89. Kant, 159, 161, 169.

90. Bonneuil and Fressoz, *Shock of the Anthropocene*, 80, emphasis added.

91. Kant, "An Old Question," 161.

92. Jacques Lacan, "On the Network of Signifiers," in *The Four Fundamental Concepts of Psycho-analysis*, trans. Alan Sheridan and ed. Jacques-Alain Miller (Harmondsworth: Penguin, 1977), 45.

93. Bonneuil and Fressoz, *Shock of the Anthropocene*, 94, 95.

94. Lacan, "The Line and Light," in *Four Fundamental Concepts*, 103.

95. Nigel Clark, "Geo-politics and the Disaster of the Anthropocene," *Sociological Review* 62, no. S1 (2014): 27–28. See also Nigel Clark, "Politics of Strata," in "Geosocial Formations and the Anthropocene," special issue, *Theory, Culture and Society* 34, no. 2/3 (2017): 1–21.

96. Sheila Jasanoff, "A New Climate for Society," *Theory, Culture and Society* 27, no. 2/3 (2010): 237.

97. For a beginning, see Nigel Clark and Yasmin Gunaratnam, "Earthing the Anthropos? From 'Socializing the Anthropocene' to Geologizing the Social," *European Journal of Social Theory* 20, no. 1 (2017): 111–31, and Bronislaw Szerszynski, "The Anthropocene Monument: On Relating Geological and Human Time," *European Journal of Social Theory* 20, no. 1 (2017): 146–63.

98. Reznick, *The "Origin,"* 311.

99. Thanks to Timothy Morton for discussions on this point.

100. Edmund Husserl, *The Crisis of European Sciences and Transcendental Phenomenology: An Introduction to Phenomenological Philosophy*, trans. David Carr (Evanston, IL: Northwestern University Press, 1970), 142–43.

101. Edmund Husserl, "The Origin of Geometry" appended to *Crisis of European Sciences*, 358.

102. Jacques Derrida, *Edmund Husserl's "Origin of Geometry": An Introduction*, trans. and with a preface by John Leavey Jr., ed. David B. Allen (New York: Nicolas Hay, 1979), 83–84.

103. Jan Zalasiewicz and Mark Williams, *The Goldilocks Planet: The Four Billion Year Story of the Earth's Climate* (Oxford: Oxford University Press, 2012).

104. Nigel Clark, *Inhuman Nature: Sociable Life on a Dynamic Planet* (London: Sage, 2011), 5. I am indebted to Clark for drawing my attention to the Husserl text I discuss here.

105. Ludwig Wittgenstein, *On Certainty*, ed. G. E. M. Anscombe and G. H. von Wright, trans. Denis Paul and G. E. M. Anscombe (1969; New York: Harper, 1972), 13e.

CHAPTER EIGHT

1. *Oxford English Dictionary*, 3rd ed., s.v. "Mutual," https://www.oed.com/view/Entry/124381.

2. Immanuel Kant, *The Critique of Judgement*, trans. James Creed Meredith (1928; Oxford: Clarendon, 1973; first published in German, 1790), 24.

3. Douglas Burnham, *An Introduction to Kant's "Critique of Judgement"* (Edinburgh: Edinburgh University Press, 2000), 31.

4. Martin Heidegger, "The Origin of the Work of Art," in *Poetry, Language, Thought*, trans. Albert Hofstadter (New York: Harper and Row, 1975), 49.

5. Walter Benjamin, "Theses on the Philosophy of History," in *Illuminations*, trans. Harry Zohn, Thesis 6 (London: Fontana/Collins, 1982; first published in German 1955), 257.

6. *Life*, like *death*, remains a tricky and undefinable word. As Eugene Thacker puts it, "Life is that which renders intelligible the living, but which in itself cannot be thought, has no existence, is not itself living." Thacker cited in Cary Wolfe, *Before the Law: Humans and Other Animals in a Biopolitical Frame* (Chicago: University of Chicago Press, 2013), 57. In a chapter called "Life Itself" in his book *In the Beginning: The Birth of the Living Universe* (London: Penguin, 1994), the astrophysicist-turned-science writer John Gribbin, remarks, "Trying to write down a definition of life is like trying to write down a definition of time. Just as we all know what time is, until someone asks us to explain it, so we all know what life is, until someone asks us to explain it" (45). The chemist Addy Pross has written a book on the subject in which the subtitle of the book serves as an answer to the question the title poses: *What is Life? How Chemistry Becomes Biology* (Oxford: Oxford University Press, 2014). See also the discussion on p. 3 and on abiogenesis and biological evolution in chap. 8. Analytical philosopher Michael Thompson meditates on the unthinkability of life in his *Life and Action: Elementary Structures of Practice and Practical Thought* (Cambridge, MA: Harvard University Press, 2008), chap. 2, "Can Life Be Given a Real Definition?"

7. I have benefited from discussing many of these points with Frédéric Worms. See his book, *Pour un humanism vital: Lettres sur la vie, la mort et le moment present* (Paris: Odile Jacob, 2019).

8. Annie Cohen Solal, *Sartre: A Life* (London: Heinemann, 1987; first published in French, 1985), 91–93, 95, 117.

9. The only exception I know to this statement is my colleague Martha Nussbaum's essay comparing Tagore's ideas with those of Auguste Comte and John Stuart Mill. See

Martha Nussbaum, "Reinventing Civil Religion: Comte, Mill, Tagore," *Victorian Studies* 54, no.1 (Autumn 2011): 7–34. One important difference between Comte and Tagore would be, I suppose, that Tagore never had any ideas about formalizing religion, even in a civil sense.

10. For all their theoretical complexities and internal differences, the theme of "the specialness of man" pervades this literature, where most arguments begin from premises in contemporary biology and make a "contrastive comparison between humans and animals" central to their development. See the excellent discussion in Joachim Fischer, "Exploring the Core Identity of Philosophical Anthropology through the Works of Max Scheler, Helmuth Plessner, and Arnold Gehlen," *Iris*, April 1, 2009, 153–70. Fischer writes, "The key concepts Scheler introduces to describe 'man's place in nature' are *Neinsagenkonner* [he who can say no], *Weltoffenheit* [openness to the world], and the ability for the living being in question to regard something as having a *Gegenstand-Sein* [to be an object]" (158–59). Notice the broad similarities with Tagore's argument. I owe this reference to Hannes Bajohr.

11. Rabindranath Tagore, "The Religion of Man," in *The English Writings of Rabindranath Tagore*, vol. 3, *Miscellany*, ed. Sisir Kumar Das (New Delhi: Sahitya Akademi, 1999), 127.

12. Tagore, 83–189.

13. Tagore, 87.

14. Tagore, 88.

15. Robin Dunbar, *Human Evolution: Our Brains and Behavior* (New York: Oxford University Press, 2016), p. 115. See also p. 98: ". . . by 4 million years ago at the latest, we have a distinctive lineage of bipedal apes in Africa, with well-established concentrations in eastern and southern Africa." See also the discussion in the section entitled "Two legs are good" in Simon L. Lewis and Mark A. Maslin, *The Human Planet: How We Created the Anthropocene* (London: Penguin Random House, 2018), pp. 82–88.

16. Tagore, "The Religion of Man," 103.

17. Tagore, 103.

18. Tagore, 103.

19. Tagore,104.

20. Tagore, 126, emphasis added.

21. Tagore, 121.

22. Tagore, 89.

23. Tagore, 104.

24. Tagore, "Lecture 1, Man," in *The English Writings of Rabindranath Tagore*, vol. 3, *Miscellany*, ed. Sisir Kumar Das (New Delhi: Sahitya Akademi, 1999), 194.

25. Rabindranath Thakur [Tagore], "Amar jagat" [My world], in his *Sanchay*, vol. 12 in *Rabindra rachanabali* [The collected works of Rabindranath], centenary edition (Calcutta: Government of West Bengal, 1961), 565.

26. Tagore, "The Religion of Man," 89.

27. Martin Hägglund, *This Life: Secular Faith and Spiritual Freedom* (New York: Pantheon Books, 2019), 173, emphasis added.

28. Hägglund, 173.

29. William James, *The Varieties of Religious Experience*, vol. 1 of *Writings* (New York: Library of America, 1987), 36, emphasis added.

30. Martin Buber, *I and Thou*, trans. Ronald Gregor Smith (Edinburgh: T. and T. Clark, 1937).

31. Bede Griffiths, *The Golden Spring* (London: Fount, 1979), 9, cited in Charles Taylor, *A Secular Age* (Cambridge, MA.: Harvard University Press, 2007), 5.

32. Søren Kierkegaard, *Papers and Journals: A Selection*, translated and introduced by Alastair Hannay (London: Penguin, 1996), 26–27, emphases added.

33. Emmanuel Levinas, *Otherwise than Being, or Beyond Essence*, trans. Alphonso Lingis (1978; Dordrecht: Kluwer Academic).

34. See the discussion in the previous chapter.

35. Edward Burtynsky, Jennifer Baichwal, and Nick de Pencier, *Anthropocene* (Göttingen: Steidl, 2018), 21.

36. See https://ia800207.us.archive.org/20/items/OnBeingTheRightSize-J.B.S.Haldane/rightsize.pdf.

37. Georges-Louis Leclerc, le comte de Buffon, *The Epochs of Nature*, trans. Jan Zalasiewicz, Anne-Sophie Milton, and Matuesz Zalasiewicz (Chicago: University of Chicago Press, 2018), 119. The medieval theological literature on the idea of the creature, as Eugene Thacker shows, is rich and vast in its complexity. I am using the word in a minimalist sense, ignoring the theological question of the relationship between a creature and its creator. "The creature is unique," writes Thacker, "because, while it may be argued that its essence is of supernatural origin, its existence takes place within the domain of nature. And this existence can only be described as 'living.'" Eugene Thacker, *After Life* (Chicago: University of Chicago Press, 2010), 106.

38. Thacker, *After Life*, xv.

39. See the discussion in Marcia Bjornerud, *Timefulness: How Thinking Like a Geologist Can Help Save the World* (Princeton, NJ: Princeton University Press, 2018), 13.

40. David Keith, *A Case for Climate Engineering* (Cambridge, MA: MIT Press, 2013).

41. Keith, *Case for Climate Engineering*, xvi–xvii.

42. Edward O. Wilson, *Naturalist* (Washington, DC: Island Press, 1994), 360.

43. Cary Wolfe, *Before the Law: Humans and Other Animals in a Biopolitical Frame* (Chicago: University of Chicago Press, 2013), 59–60.

44. Andrew Y. Glikson and Colin Groves, *Climate, Fire and Human Evolution: The Deep Time Dimensions of the Anthropocene* (London: Springer, 2016), 194. On the question of reverence for Terra as a mother figure in European history, see the discussion in Philip John Usher, *Exterranean: Extraction in the Humanist Anthropocene* (New York: Fordham University Press, 2019), chap. 1.

45. Bjornerud, *Timefulness*, 157.

46. Wallace S. Broecker and Robert Kunzig, *Fixing Climate: What Past Climate Changes Reveal about the Current Threat—and How to Counter It* (New York: Hill and Wang, 2008), 100.

47. Edward O. Wilson, *Half-Earth: Our Planet's Fight for Life* (New York: W. W. Norton, 2016), 89, 194–95.

48. Martin Heidegger, "Building Dwelling Thinking," in his *Poetry, Language, Thought*, trans. Albert Hofstadter (New York: Harper and Row, 1971), 148–49.

49. Robin Dunbar, *Human Evolution: Our Brains and Behavior* (New York: Oxford University Press, 2016), 125–26.

50. On the story of the dingo, see Deborah Bird Rose, *Dingo Makes Us Human: Life*

and Land in an Australian Aboriginal Culture (Cambridge: Cambridge University Press, 2000).

51. An elaborated instance of what I have described as thinking from a minority position is Faisal Devji, *The Impossible Indian: Gandhi and the Temptation of Violence* (Cambridge, MA: Harvard University Press, 2012).

52. K. Jaspers, *Man in the Modern Age*, trans. Eden and Cedar Paul (New York: Henry Holt, 1933; first published in German, 1931), 1. Paul Valery's famous two 1919 letters, *Crisis of the Mind* (*La crise de l'esprit*) come to mind here. See Edward J. Hundert, "Oswald Spengler: History and Metaphor—The Decline and the West," *Mosaic: An Interdisciplinary Critical Journal* 1, no. 1 (October 1967), 103–17.

53. Jaspers, *Man in the Modern Age*, 4.

54. Jaspers, 5–6.

55. Jaspers, 6.

56. Jaspers, 7–8.

57. Jaspers, 8–16.

58. A case in point is Jaspers's own book *The Atom Bomb and the Future of Man*, trans. E. B. Ashton (Chicago: University of Chicago Press, 1963). An earlier 1961 edition was published under the title *The Future of Mankind*. The original German edition was published in 1958.

59. Jaspers, *Atom Bomb*, 10, 12–13.

60. Jaspers, 222, 223, 307.

61. Andrew Y. Glikson and Colin Groves, *Climate, Fire and Human Evolution: The Deep Time Dimensions of the Anthropocene* (London: Springer, 2016), 194.

62. *Oxford English Dictionary*, 3rd ed., s.v. "reverence," https://www.oed.com/view/Entry/164755#eid25518919; *Etymological Dictionary of Latin*, ed. Michiel de Vaan (Leiden: Brill, 2010), s.v. "vereor," https://dictionaries.brillonline.com/search#dictionary=latin&id=la1800.

63. See the discussion in Rudolf Otto, *The Idea of the Holy: An Inquiry into the Non-rational Factor in the Idea of the Divine and Its Relation to the Rational*, trans. John W. Harvey (Mansfield Center, CT: Martino, 2010; first published in English 1923 from *Das Heilige* [1917]), chaps. 4, 5, and pp. 20–21, 24. I am grateful to David Lamberth and Charles Hallisey for encouraging me to read Otto. Otto translates "tremendum mysterium" as "awful mystery" (25).

64. John Locke, *Second Treatise of Government*, ed. and with an introduction by C. B. Macpherson (1690; Indianapolis, IN: Hackett, 1980), 21.

65. Hugo Grotius, *Mare liberum*, ed. and annotated by Robert Feenstra (Leiden: Brill, 2009; first published in Latin, 1609), 121. See also the discussion in Davor Vidas, "Oceans in the Anthropocene—and Rules of the Holocene," in *Welcome to the Anthropocene: The Earth in Our Hands*, ed. Nina Möllers, Christian Schwägerl, and Helmuth Trischler (Munich: Deutsches Museum, 2015), 56–59.

66. Jeffrey H. Schwartz, "What Constitutes *Homo sapiens*? Morphology versus Received Wisdom," *Journal of Anthropological Sciences* 94 (2016): 1–16; Bernard Wood and Mark Collard, "The Meaning of Homo," *Ludus vitalis* 9, no. 15 (2001): 63–74.

67. See chapter 6 for the reference to Kant. On the origins of "cornucopian" thoughts in Europe, see Fredrik Albritton Jonsson, "The Origins of Cornucopianism: A Preliminary Genealogy," *Critical Historical Studies* 1, no. 1 (2014): 1–18.

68. See the discussion on "Multinaturalism" in Eduardo Viveiros de Castro, *Cannibal Metaphysics*, trans. and ed. Peter Skefish (Minneapolis, MN: Universal, 2014; first published in French, 2009), 65–75, and the chapter on "Animism Restored" in Philippe Descola, *Beyond Nature and Culture*, trans. Janet Lloyd (Chicago: University of Chicago Press, 2013; first published in French, 2005), 129–43.

69. Theodore Adorno and Max Horkheimer, *Dialectic of Enlightenment*, trans. John Cumming (London: Verso, 1979; first published in German, 1944), 3.

70. Annu Jalais, *Forest of Tigers: People, Politics, and the Environment in the Sundarbans* (New Delhi: Routledge, 2010), and Vijaya Raghavan, *Feeding a Thousand Souls: Women, Ritual, and Ecology in India — An Exploration of the Kōlam* (New York: Oxford University Press, 2019), chaps. 9, 10. Ironically, it has to be noted, the fear of wild animals is coming back in certain parts of big cities in India today as monkeys, leopards, elephants, and even rhinoceroses, deprived of their habitats by environmental crises and expansion of human settlements, are driven to the latter in search of food and water.

71. Referring to the distinction between wild and domesticated, Descola remarks, "it is unlikely that [this] . . . distinction can have been at all meaningful in the period prior to the Neolithic revolution — that is to say, during the greater part of human history." Descola, *Beyond Nature and Culture*, 35.

72. Thomas Hobbes, *On the Citizen*, ed. and trans. Richard Tuck and Michael Silverthrone (1998; Cambridge: at the University Press, 2006; first published in Latin, 1642; 2nd ed. 1647), 105.

73. Hobbes, *On the Citizen*, 106.

74. Thomas Hobbes, *Leviathan*, ed. and with an introduction by C. B. Macpherson (1651; Harmondsworth: Penguin, 1976)), 116–17.

75. Hobbes, *Leviathan*, chap. 5, pt. 1, 117.

76. Hobbes, 116–17.

77. Hobbes, chap. 3, pt. 1, 98.

78. In a powerful recent discussion of Gandhi's ethics, Ajay Skaria has explained what Gandhi meant by "fearlessness" while leading his people in their struggle for freedom from colonial rule. "*Abhay*, the Gujarati word he [Gandhi] translates as both courage and fearlessness, is a crucial term in Gandhi's vocabulary," writes Skaria. It was not about building a human-dominant order: "It means a different kind of courage — not physical but moral courage. Above all, moral courage involves questioning oneself, reflecting on whether one's actions are right or wrong. . . . To be internally divided is to develop a conscience, to become capable of having an interminable conversation with oneself about right and wrong, beginning with the right and wrong of one's own actions." Ajay Skaria, "Gandhi and the Cowardice of Hindutva," *Wire* (India), October 2, 2019, https://thewire.in/history/gandhi-and-the-cowardice-of-hindutva.

79. This is not only the position of those of who favor geoengineering options; it has been held more widely. One comes across it, for example, in what was written about animal rights and welfare before global warming was seen as a critical and unavoidable issue. See Nussbaum's discussion — grounded in Aristotle's theory of wonderment — of the question of justice for animals in Martha C. Nussbaum, *Frontiers of Justice: Disability, Nationality, and Species Membership* (Cambridge, MA: Harvard University Press, 2006), 348–52.

80. Roy Scranton, *Learning to Die in the Anthropocene: Reflections on the End of a Civilization* (San Francisco: City Light Books, 2015).

POSTSCRIPT

1. See chap. 15, "Machinery and Modern Industry," in Karl Marx, *Capital*, vol. 1, bk. 1 *The Process of Production of Capital*, trans. Ben Fowkes (London: Penguin, 1990; first published in German in 1867), 492–639.

2. Carl Schmitt, *Dialogues on Power and Space*, ed. Andreas Kalyvas and Federico Finchelstein, trans. Samuel Garrett Zeitlin (Cambridge: Polity Press, 2015; first published in German, 1954).

3. Bill McGuire, *Waking the Giant: How a Changing Climate Triggers Earthquakes, Tsunamis, and Volcanoes* (New York: Oxford University Press, 2012).

4. Rachel Carson, *Silent Spring* (Boston: Houghton Mifflin, 1962); Donella H. Meadows et al., *The Limits to Growth: A Report for the Club of Rome's Project on the Predicament of Mankind* (New York: Universe Books, 1972).

5. Fernand Braudel, *On History*, trans. Sarah Matthews (Chicago: University of Chicago Press, 1980; first published in French, 1969), 10.

6. Theodor W. Adorno and Max Horkheimer, *Dialectic of Enlightenment* (New York: Herder and Herder, 1972; first published in German, 1947).

Index

academic disciplines, and climate change, 38
acidification of oceans. *See* ocean acidification
action, in Arendt, 8–10, 229n24, 229n28, 230n32
Adorno, Theodor W.: on fear, 217; negative universal history in, 45, 46–47
Agarwal, Anita, 96, 159–60, 232n61
agency: geological, 3; historical, 3; planetary, 4
agricultural revolution, 39, 42; geophysical parameters of, 40–41
air conditioning, and climate change, 97, 98–99
Althusser, Louis: and Marx, 234n16; on concepts, 43
Angus, Ian, 161
animals: competition with humans, 90, 126; habitat pressure, 59–60; and politics, 13
Anthropocene: academic literature on, 227n5; and deep time, 228n19; defined, 3; embrace of, 162–64; as geological label, 3, 6, 13, 33, 158, 169–70, 227n7; history of, 42; history of the term, 156–57; and human geophysical power, 157; and humanities, 14, 40, 155, 227n5, 236n42; as idea/concept, 3, 7, 14, 33; and industrialization, 40; naturalized, 208; phenomenologically inaccessible to humans, 79; and political thought, 228n23; popular literature on, 182; popular reception of, 158; possible beginning dates of, 33, 39, 61–62, 155, 167–68; skepticism of, 16, 155, 160–61, 176; "social," 6; public discussion of, 1; unintended, 34
anthropocentrism, 62–65; multiple, 63; religious perspectives on, 63–64; strong and weak, 64–65
anthropology, new philosophical, 20, 90, 91
anticolonial nationalism, and modernization, 106
Archer, David: on carbon cycle, 66; on human climate agency, 50; on human exceptionalism, 63; on human timescales, 51; *Long Thaw, The*, 162
Arendt, Hannah: on action, 8–10, 229n24, 229n28, 230n32, and Benjamin, 229n26; and Heidegger, 229n26; *Human Condition, The*, 8–9, 71–72; on labor, 9, 10, 205–6, 230n32; *Life of the Mind, The*, 9; on the political, 8–10; *Promise of Politics, The*, 9; and Sputnik, 80; and technology, 80; on work, 9, 205–6, 229n24, 230nn32–34, 230n36
Aristotle, 134, 217
Arrhenius, Svante, 24, 233n3
Arrighi, Giovanni, 25, 237n66
astrobiology, 173

atmosphere, dominated by humanity, 8
Augustine, Saint, 164, 171
Australian Capital Territory, 2

bacteria: anaerobic, 87; history of, 150
Balibar, Étienne, 234n16
Baucom, Ian, 231n50
Bayly, C. A., 24
Bedford, Ian, 234n16
Benjamin, Walter, 183, 229n26
Bennett, Jane, 13, 20, 90, 91, 184, 212. *See also* thin description
Bergthaller, Hannes, 14
Berlant, Lauren, 236n37
Bhabha, Homi K., 17
biodiversity, 42; loss of, 7, 34, 36, 84, 91, 127; in public discussion, 1; wonder at, 198, 212
biology, political uses of, 37–38
biosphere, 206; dominated by humanity, 8, 62
Birmingham, Peg, 230n34
Bjornerud, Marcia, 194
"blue marble" (picture of Earth), 17, 245n2
Bonneuil, Christophe, 175–77
Brahmin, sedentary lifestyle, 15
brain, human, 138, 228n22
Braudel, Ferdinand, 215; environmental cycles in, 29
Bright, Charles, on humanity, 44; on multiple modernities, 37
Broecker, Wally [Wallace], 83–86, 88, 171, 172, 194
Brooke, John L., 164
Brundtland Commission. *See* World Commission on Environment and Development
Buffon, Comte de (Georges-Louis Leclerc), 171, 191–92
Bush, George H. W., 25
bushfires. *See* wildfires

Calcutta, 2
calendar of political action on climate change, 12–13
Canberra, 2
"Can the Subaltern Speak?" (Spivak), 17, 232n57

capitalism: against logic of dwelling, 9–10; and Anthropocene, 39–40, 41–42; and consumption, 231n41; critiques of, 35; and globe and planet, 4; insufficient to explain Anthropocene, 4, 35, 51
Capitalocene, 16, 101, 155
carbon cycle, 90, 128
carbon sinks, as global commons, 58
Carson, Rachel, 17, 202, 208
caste, historiography of, 118–19
Catton, William, Jr., 139
Césaire, Aimé, 106
China, 41; coal use in, 55, 59
chronologies, multiple, 14–15. *See also* timescales
civilization, 133–34
Clark, Brett, 46, 56
Clark, Nigel, 177–78, 180
climate change: and climate justice, 4, 56; crises of, 3; as "dangerous," 12, 13, 133, 158; discussion of in the West, 234n11; as greatest ethical challenge, 146; and industrialization, 24; intended and unintended, 45; intensifies inequality, 11–12, 35, 45; narratives of, 135–7; popular writings on, 24, 133, 233n9, 234n11; and postcolonial thought, 24; probabilistic descriptions of, 52, 53–54; in public discussion, 1, 24; scientific consensus on, 25–26, 234n11; and sense of the now, 23; signs of, 25; and sovereignty, 95; uncertainties of, 54–55; vitalist descriptions of, 52–53; writings on, 25–26
climate justice: and capitalism, 4; and carbon space, 73; as political, 10; challenge to ideas of human, 24; developed vs developing countries, 41
"Climate of History, The: Four Theses" (Chakrabarty), 14–15, 18, 23–48; responses to, 15–16, 43, 46, 231n50, 233n7, 239n86
Club of Rome, 17. See *Limits to Growth, The*
coal, as fossil fuel, 59
Collingwood, R. G., on history, 43; on natural and human history, 25–28; as translator of Croce, 27, 234n14, 235n17
common but differentiated responsibility, 41, 56–57

commons, new, 20
Connolly, William, 19, 85–86
cosmopolitanism, 46
cost-benefit analysis, 54
COVID-19, 1
critical zone, 69, 77
Croce, Benedetto: idealism in, 28; on natural and human history, 26–27; on Vico, 26–27, 235n18
Crosby, Alfred, Jr., 30–31
Crutzen, Paul J., 32–33; and origins of the Anthropocene, 156–57; on great acceleration, 61; on humans as a species, 37

Dalit, as caste identity, 19; multiplicities of, 122–23
Dalit body, 19, 117–19, 122
Darwin, Charles, 171
Davies, Jeremy, 14, 228n19, 235n30
Davis, Mike, on periodization, 33; on population, 34, 98
deep earth, 206
deep history, 8, 36, 138; and multiple timescales, 15; politically challenging, 151
deep time, 8, 183
Derrida, Jacques, on madness, 38, 44, 179–80
Descola, Philippe, 105
developing countries, aspirations for modernization in, 98–99, 100–101
"Dialogue on New Space" (Schmitt), 6
Dilthey, Wilhelm, on historical understanding, 43, 45
disgust, of the body, 115–16
diversity, and freedom, 31
domination, human, 44
Dumont, Louis, 123–24, 125
Dunbar, Robin, 195
durability, 9, 230n33
Durkheim, Émile, 63–64
Dutreuil, Sébastien, 216
dwelling, in Arendt, 9

Earth (capitalized), as abstract and unvisualizable entity, 245n2
earth (lowercase), as organizing category of modern history, 3; in Heidegger, 183, 206; in relation to humans, 70, 183

Earth system, 4, 70, 77, 172. *See also* planet
Earth System Science, 18, 68, 172, 209; as comparative science, 67, 75, 79; defined, 75, 76; elements of, 77–78; and Gaia, 80; and the humanities, 14; origins of, 75–76, 78; as view from outside, 78
elites, anticolonial and modernizing, 2
empire, 2; alongside biological history, 7–8; and indigenous sovereignty, 10
energy: extraction of, 5; increases in use of, 4
environmental colonialism, 96, 159
environmental degradation, examples of, 91, 95
environmental ethics, 62–65
environmental history, 29–30
environmentalism, growth of, 17
environmental management, 34. *See also* geoengineering
epochal consciousness, 19
ethics, Anthropocene, 62–65
Europe: and empire, 2; and modernizing elites, 2
evolutionary science, 30
extinction, 25, 31, 40, 55; human, 23, 45
Exxon, 34, 50, 234n45

Fanon, Franz, criticism on, 17, 123
Flannery, Tim, 35, 133
food chain, 42
forests, 5
fossil fuel: and democracies, 237n52; and industrial civilization, 66; public discussion of, 1; use of, 32, 40; as renewable on long timescales, 49, 85
Foster, John Bellamy, 46, 56, 156
Foucault, Michel, 38, 44
Francis I (pope), 134, 146–47
freedom: as energy intensive, 32; and geological agency, 32; as historical motif, 32; modern ideas/concepts of, 19
Fressoz, Jean-Baptiste, 175–77
futures, alternative, 35

Gadamer, Hans Georg: on Dilthey and history, 43; on diversity and freedom, 31; on earth and planet, 69; on prejudice, 120; on technology, 80

Gaia, 78, 80, 209, 212
Gandhi, Mahatma, 59, 103, 202
Gardiner, Stephen, 230n37
geoengineering, 175; arguments for, 87–88, 89; disciplinary predispositions toward, 192; limits on, 90; literature on, 230n39; as political, 10; political consequences of, 12; problems of, 35
geological events, in human history, 7
Geological Society of London, 33, 49
geologists, methodology of, 166–67
Geyer, Michael, 37, 44
Ghosh, Amitav, 59
glacial-interglacial cycle, 11
Glikson, Andrew, 174–75, 194, 198, 199, 202
global, age of the, 182
globalization, and history of life, 1. *See also* capitalism; globe
global warming, 3, 42; average surface temperature rise, 133, 136; and capitalism, 4; origins of scientific study of, 24, 233n3; as a public concern, 24
globe: as anthropocentric concept, 4, 75; different meanings of, 18; as organizing category of modern history, 3
globe/planet distinction, 18, 51, 69–74. *See also* globe; planet
God species, 89, 92, 147, 175
great acceleration, 61–62, 139
great extinction, sixth, 7, 31, 36, 84, 127, 140, 178, 183
great oxygenation event, 166, 173
greenhouse effect, as necessary for life, 128
greenhouse gas, 4, 24, 128
Griffiths, Bede, 189
Grinspoon, David, 162–63
Groves, Colin, 174–75
Guha, Ranajit, 3, 98, 118

habitability, 83–85, 87; and humans, 83–84
Haff, Peter, 5, 173
Hägglund, Martin, 188–90
Haldane, J. B. S., 191
Hall, Stuart, 17
Hamilton, Clive, 163
Hansen, James, 16; on climate change, 24; on coal, 55; as planetary scientist, 75

Harari, Yuval Noah, 138–39
Haraway, Donna, 13
Hartog, François: on "end-time," 231n40; on regimes of historicity, 245n1
Hegel, G. F. W., 1; on freedom, 31; on state, 70; universals of, 45
Heidegger: on dwelling, 195; on earth and planet, 68–69; on planet, 74; on technology, 80
Heise, Ursula: on human species, 43–44, 238n81; on species and universals, 46; on Žižek, 66
historical consciousness, 43
historicism, paradoxes of produced by climate change, 23
histories, juxtaposition of, 20, 49
history: assumptions underlying, 23, 39; collapse of distinction between natural and human, 30; deep, 7, 8, 36, 138; as humanist discipline, 23–24; natural vs. human, 26–31; recorded, 7, 36; shared among species, 228n22; as story of creation, 234n12
history of industrial civilization, 49
history of life, 1, 42, 49
history of the Earth system, 49
Hobbes, Thomas, 71, 200–202
Holocene, 3, 7, 32, 33
Horkheimer, Max, 217
Horn, Eva, 14
Hornborg, Alf, 160–61
How to Build a Habitable Planet (Langmuir and Broecker), 172, 173, 174
human being, varied academic views of, 38
human capital, 112
Human Condition, The (Arendt), 8–9, 20
human dominance, 136
human evolution, 137, 138–39
human geophysical force, seen as an experiment, 158
humanities, history of, 16–17, 232n54
humanity, 9, 14–15; ecological overshoot of, 137–38; polarized and divided, 37; rise to top of food chain, 137, 138. *See also* humans
humans: animal life of, 134–35; as biological agent, 7, 30–31; divided, 14, 37, 51; effects of on climate, 41–42; as geologi-

cal/geophysical agent, 3, 6–7, 11, 14, 30–31, 32; in humanist history, 3; moral life of, 134–35; as a species, 36–38, 51. *See also* humanity
Husserl, Edmund, 179–80
hyperobjects, 79

ice-core samples, 49–50
India: Anthropocene in, 11; coal use in, 55, 59; and globalization, 112; and greenhouse gases, 112; population growth in, 40
industrialization, and Anthropocene, 40
Industrial Revolution, and human geophysical agency, 31
inequality, between classes, 10
intergenerational politics, 10, 230n37
Intergovernmental Panel on Climate Change (IPCC), 13, 16, 214; challenges to, 89, 96; establishment of, 16; reports of, 25, 159
irrigation, 231n49

Jasanoff, Sheila, 159
Jaspers, Karl, 184; on epochal consciousness, 19, 48, 196–98; on technology, 80
Jetztzeit, 7
Johnson, Harriet, 47, 239n88
Jonas, Hans, 135

Kant, Immanuel: on anthropology, 20; *Critique of Judgment*, 183; on domains of knowledge, 20; on freedom, 31; on Genesis, 141–45; on moral and animal lives of humans, 19, 142, 145–46; and natality, 229n26; on progress, 175–76; on Rousseau, 142; on wonder, 115
Keith, David, 193, 194, 213
Kelly, Duncan, 228n23
Kierkegaard, Søren, 189–90
Koselleck, Reinhart, 164–66

labor, 9, 10, 205–6, 230n32
landfills, 6
Langmuir, Charles H., 83, 84, 85, 86, 88, 171, 172
Larkin, Peter Anthony, 82
Latour, Bruno, 103–5; on the commons, 194; on the "Constitution of the Modern," 103–4; in conversation with Dipesh Chakrabarty, 20, 205–17; on Earth System Science, 76–77; on human dominion, 151; on Kant, 149, 150–51; on microbes, 149–50; on nature-human distinction, 236n33; on the parliament of things, 100; on politics beyond the human, 13, 239n94; on rebarbarization, 91, 230n39; on the sciences, 35
LeCain, Timothy J., 238n80
Leibniz, Gottfried Wilhelm, 70
Lenton, Tim, 76–77, 84, 88, 216
Lewis, Simon L., 167–68
life, history of, 1
Limits to Growth, The (Meadows et al.), 17, 208
Lisbon earthquake (1755), 70
lithosphere, dominated by humanity, 8
local, the, and the Anthropocene, 14, 231n46
Locke, John, 199
Lovell, Bryan, 49, 50
Lovelock, James, and Gaia, 75, 78
lumpers and splitters, 17–18
Lyell, Sir Charles, 33

Macpherson, C. B., 15
Malabou, Catherine, 228n22
Malm, Andreas, 160–61
Man, question of, 184
Markell, Patchen, 13, 229n24, 230nn32–34, 230n36,
Marx, Karl: on capital, 70; on freedom, 31; on history, 27
Marxism, responses to climate change, 135
Maslin, Mark A., 35, 167–68
maximin principle, 55
McGuire, Bill, 207
McKibben, Bill, 236n32
McNeill, John, 4, 148; on the great acceleration, 61
McNeill, William, on human geophysical capacity, 163–64
Meadows, Dennis, 17
Meadows, Donella, 17
microbes, 117, 125–26, 134, 193–94; as majority of biomass, 148

migrations, animal and human, 60
Milankovitch cycles, 40–41, 171
Mitchell, Timothy, 237n52
modernity, multiple, 37
modernization: and greenhouse gases, 4; hope for in developing countries, 98–99, 101–2; and industrialization, 4
modern political thought, 211–12
Montreal Protocol, 12, 214
Moore, Jason, 46, 161
Morrison, Kathleen D., 167, 231n49
Morton, Timothy, 79
mutuality, 19–20, 182–83, 187–91

Narain, Sunita, 56, 96, 159–60, 232n61
natality, in Arendt, 8–9, 229n24, 229n26, 230n32
negative universal history, 45, 46–48
Nehru, Jawaharlal: and irrigation, 106–10; and modernization, 106–10, 111–12
nervous system, shared by humans and other animals, 228n22
neuroscience, 30
Nixon, Rob, 232n61
nonanthropocentric thought, 51
nonhuman life, and politics, 13
Nussbaum, Martha, 103; on disgust, 115–17

ocean acidification, 11, 42, 60, 91
ocean levels, 42
Osborn, Fairfield, 82
overshoot, 138–40
oxygen, levels of, 5, 79, 84, 117

Paleocene-Eocene Thermal Maximum (PETM), 50, 191; and widespread extinction, 55
Pettit, Philip, 129–30
philosophical anthropology, 20, 90, 91, 184
planet, 1, 3, 68–71; as challenge to political thought, 91; decenters the human, 4; deep, 77; differs from globe, 18–19; as Earth system, 70; human agency in, 4; human effects on, 10–11; revealed by globalization, 69, 80, 182; undermines mutuality, 182; and view from outside, 78–79

planetary, the, 3, 85. *See also* planet
planetary boundaries, 139–40, 238n69
planetary science, 67, 75. *See also* Earth System Science
planeticide, 175
plankton, 5
Pleistocene, 32
plurality, in Arendt, 8–10
politics: beyond the human, 13; indispensability of, 231n41; multispecies, 44; nonanthropocentric, 151; short-term, 35
Politics of Nature (Latour), 35
Polyani, Karl, 61
Pomeranz, Kenneth, 39–40
population, 58–61; and deep history, 60–61; growth, 34–35, 41, 98; growth in developing countries, 98; growth in twentieth century, 4, 25, 40; reduction, 149
postcolonial thought: and Anthropocene, 10; and climate change, 24; and environmentalism, 17; origins of, 16–17; suspicion of universals in, 42
precautionary principle, 55
prehistory, 36
prejudice, 120–22
probabilistic thought, 51, 52
Provincializing Europe: Postcolonial Thought and Historical Difference (Chakrabarty), 2, 13, 48
purification, and commodification, 150

Raffinsøe, Sverre, on Kant, 20
Rajasthan, geomorphological role in, 11, 230n38
Ramachandran, Ayesha, 78–79
reason, in politics, 34
regime of historicity, 68, 96
renewable energy, as topic of public discussion, 1
reverence, 198, 212
Reznick, David, 178–79
risk, 52
Roberts, David, on Croce, 28, 235n17
Ruthven, Kenneth, 2

Sachs, Jeffrey, 38
Sagan, Carl, 114, 116

Said, Edward, 16
Schmitt, Carl: on dwelling, 71; on land and territory, 72–73; on law/*nomos*, 72–73; on planet and globe, 73–74; on the political, 8, 13, 231n41; on the sea, 72; on technology, 5–6, 80
Schrag, Daniel, 163
science: and politics, 14, 231n45; skepticism of, 14, 231n45
science studies, 14, 231n44
scientists, use of Enlightenment thought by, 42
Scott, James C., 16–17
Scranton, Roy, 203–4
sea-bottom trawling, 7
Sen, Amartya, 32, 103; *Development as Freedom*, 112; on energy use and development, 147; on human responsibility to nonhuman, 147–48
Sha Zukang, 56
Shiva, Vandana, 17
Singer, Peter, on climate change, 134
Smail, Daniel Lord, 30, 228n22; on biology and politics, 37–38; on species, 38
socialism, 40
social media, 10
Something New Under the Sun (McNeill), 158
species, 42; humanity as, 36–38, 43; objections to the term, 39; term as used by scholars, 38
species-being, in Arendt, 9
Species with Energy-Intensive Technology (SWEIT), 173
Spengler, Oswald, 80
Spinoza, Baruch, 1, 43, 238n77
Spivak, Gayatri Chakravorty, 17, 67
Stalin, Josef, on the environment, 28–29
Stamos, David N., 45
Steffen, Will, on Earth System Science, 76; on great acceleration, 61
Stengers, Isabelle, 13
Stoermer, Eugene F., 32–33
Subaltern Studies (Guha), 2, 17, 232n56; and Dalit question, 118–20
subject, Lockean, 210
Sunstein, Cass, 55
sustainability: as anthropocentric, 82–83; defined, 81

Tagore, Rabindranath, 113, 183–84, 184–88, 202; on connection with other life, 130–32
Taylor, Charles, 189
technological civilization, lifespan of, 88
technology, 5; as condition for biology, 5; encumbered and unencumbered, 6, 206–7
technosphere, 173; agency of, 6; defined, 5–6; dominant, 44; and human population, 5, 6; human role in, 14
Thacker, Eugene, 192
thin description, 100, 102–3, 113
Thomas, Julia Adeney, 133
Thompson, E. P., 3, 43
thought, planetary, 19
timescales: of biology, 7; of deep history, 4, 7; of geology, 7, 10; of humanist history, 4, 7; of human lives, 11; mixing of, 4, 7, 10–11, 32, 36, 49–51, 156, 162; of politics, 50
tipping point, 29, 54–55
twentieth century, as time of transformation, 4

unborn, politics of, 135
uncertainty, 52
United Nations: and climate change, 12–13; and planetary problems, 214; working calendar of, 214
universalism, suspicions of, 18
universals, 45
University of Chicago, 2
University of Melbourne, 2

Vanderheiden, Steve, 128–29
Vatter, Miguel, 229n24, 229n26
Vemula, Rohith, 114–15, 127, 183–84
Vico, Giambattista, on natural and human history, 26–27
Villa, Dana, 229n26

Warde, Paul, 81
water shortages, 12
Watson, Andrew, on Earth System Science, 76–77
weather, extreme, 1, 11, 12
Weisman, Alan, 23, 44–45

wicked problem, 141
wildfires, 2, 45, 90, 203
Williams, Bernard, 14; and Latour, 231n44
Williams, Mark, 6
Wilson, Edward O.: on biodiversity, 34; and biophilia, 193; on connections between biology and culture, 141; on deep history, 36; on humans as a species, 37, 38–39; on geoengineering, 194; and politics, 35
Wittgenstein, Ludwig, 180–81, 215
Wolfe, Cary, 193–94
Wood, Gillen D'Arcy, 16, 232n51
work, 9, 205–6, 229n24, 230nn32–34
Working Group on the Anthropocene, 3
world: in Heidegger, 245n12; as organizing category of modern history, 3; as phenomenological category, 8
World Commission on Environment and Development, 81
world parliament, 44
Worster, Donald, 82

York, Richard, 46, 56

Zalasiewicz, Jan, 91, 155–56; on Anthropocene, 13, 140, 141, 227n5, 227n7; on multiple Anthropocenes, 155; on planetary thinking/thought, 19
Žižek, Slavoj, 65–66
zoe, 134

Lightning Source UK Ltd.
Milton Keynes UK
UKHW021819070721
386790UK00005B/19